DNA Replication:
Procedures and Applications

DNA Replication:
Procedures and Applications

Edited by **Tom Lee**

New York

Published by Callisto Reference,
106 Park Avenue, Suite 200,
New York, NY 10016, USA
www.callistoreference.com

DNA Replication: Procedures and Applications
Edited by Tom Lee

International Standard Book Number: 978-1-63239-153-7 (Hardback)

The publisher's policy is to use permanent paper from mills that operate a sustainable forestry policy. Furthermore, the publisher ensures that the text paper and cover boards used have met acceptable environmental accreditation standards.

Trademark Notice: Registered trademark of products or corporate names are used only for explanation and identification without intent to infringe.

Printed in the United States of America.

Contents

Preface

I am honored to present to you this unique book which encompasses the most up-to-date data in the field. I was extremely pleased to get this opportunity of editing the work of experts from across the globe. I have also written papers in this field and researched the various aspects revolving around the progress of the discipline. I have tried to unify my knowledge along with that of stalwarts from every corner of the world, to produce a text which not only benefits the readers but also facilitates the growth of the field.

This book outlines and reviews the present condition of knowledge on numerous crucial aspects of the DNA replication procedure. DNA replication is a basic part of the life cycle of all the organisms. Many characteristics of this process show thorough conservation across organisms in all domains of life. DNA replication is a crucial process in both development and growth and in relation to a wide variety of pathological conditions including cancer. The book contains novel insights into the whole process of DNA replication under these sections: "Replication of Organellar Chromosome", "Chromatin and Epigenetic Influences" and "Telomeres". The topics provide the basis for thought provoking questions and summaries to give way to future investigations.

Finally, I would like to thank all the contributing authors for their valuable time and contributions. This book would not have been possible without their efforts. I would also like to thank my friends and family for their constant support.

Editor

Replication of Organellar Chromosomes

The Plant and Protist Organellar DNA Replication Enzyme POP Showing Up in Place of DNA Polymerase Gamma May Be a Suitable Antiprotozoal Drug Target

Takashi Moriyama and Naoki Sato

Additional information is available at the end of the chapter

1. Introduction

Mitochondria and plastids are eukaryotic organelles that possess their own genomes. The existence of organellar genomes is explained by the endosymbiotic theory [1], which holds that mitochondria and plastids originated from α-proteobacteria-like and cyanobacteria-like organisms, respectively [2,3]. Organellar genomes are duplicated by the replication machinery, including DNA polymerase, of the each organelle. The enzymes involved in the replication of organellar genomes are thought to be encoded by the nuclear genome and transported to the organelles after synthesis [4].

DNA polymerase γ (Polγ) is the enzyme responsible for replicating the mitochondrial genome in fungi and animals [5,6]. Polγ belongs to family A DNA polymerases, which share sequence similarity to DNA polymerase I (PolI) of *Escherichia coli*. Animal Polγ consists of two subunits: a large subunit with DNA polymerase and 3'-5' exonuclease activities, and a small subunit that enhances processivity and primer recognition. The activity of Polγ is inhibited by N-ethylmaleimide (NEM) and dideoxy nucleotide triphosphate (ddNTP).

In the late half of the 1960s, the presence of organellar DNA polymerase was confirmed by the measurement of DNA synthesis activity in isolated plant chloroplasts [7,8] and mitochondria of yeast and animals [9,10]. Since the 1970s, DNA polymerases have been purified from the chloroplasts and mitochondria of various photosynthetic organisms (Table 1), with biochemical data suggesting that plant organellar DNA polymerases and γ-type DNA polymerases share similarities with respect to optimal enzymatic conditions, resistance to aphidicolin (an inhibitor of DNA polymerase α, δ, and ε), sensitivity to NEM, molecular size, and template preference. Despite such observation, no gene encoding a homolog of Polγ has been found in the sequenced genomes of bikonts, including plants and protists. Therefore, the DNA polymerase of both mitochondria and plastids in photosynthetic organisms had remained unidentified. Sakai and colleagues [11-13] isolated nucleoid-enriched fractions from chloroplasts and mitochondria of tobacco leaves. They detected DNA synthetic activity in the nucleoid fraction and showed that the apparent molecular mass of the polypeptide exhibiting the activity was similar to Klenow fragment of DNA polymerase I (PolI) in *E. coli*. After their suggestion, it was found that the genomes of bikonts, consisting of plants and protists, encode one or two copies of genes encoding a DNA polymerase having distant homology to *E. coli* PolI. Homologs of this polymerase have been isolated in several plants, algae, and ciliates. Because genes encoding this type of enzyme are present in both photosynthetic eukaryotes and protists, we proposed to call this type of DNA polymerase POP (plant and protist organellar DNA polymerase).

| Year | Organism (organelle) | Mr (kDa) | Optimal condition for enzymatic activity | | | Inhibition by NEM (mM) | 3'-5' Exonuclease activity |
			pH	MgCl$_2$ (mM)	NaCl or KCl (mM)		
1973	*Euglena gracilis* (cp)[a]		7.2	6	10-15		
1979	Wheat (mt)[b]	110[m]	7	5	150	5	yes
1980	Cauliflower (mt)[c]				150	1	
1980	Spinach (cp)[d]	105[n]	8-9	0.1-1	100	2	
1981	Wheat (mt)[e]	180[m]	8			no	
1984	Pea (cp)[f]	87[m]		12	120	1	no
1990	Soybean (cp & mt)[g]	85-90[n]	8		125	strongly	
1991	Spinach (cp)[h]	105[n]				1	yes
1991	*Chlamydomonas* (cp)[i]	110[n]			100	2	no
1993	*Chenopodium* (mt)[j]	80-90[n]		10	125	1	yes
1995	Soybean (cp)[k]						yes
2002	Pea (cp)[l]	70[n]	7.5	8	125	partially	yes

Table 1. Previous studies on organellar DNA polymerases with no gene identification in plants and algae. cp, chloroplast; mt, mitochondrion; NEM, *N*-ethylmaleimide. a-l: references [14-25]. [m]determined by gel filtration; [n]determined by glycerol density gradient. Reproduced from [26].

2. Enzymatic characteristics of POPs

The isolation of POP was first reported in rice (*Oryza sativa*) [27,28] and later in several higher plants and algae, including thale cress (*Arabidopsis thaliana*) [29,30], tobacco (*Nicotiana tabacum*) [31], red alga (*Cyanidioschyzon merolae*) [32], and a ciliate (*Tetrahymena thermophila*)

[33]. POPs typically consist of 900-1050 amino acid residues and contain 3'-5' exonuclease and DNA polymerase domains (Figure 1). In addition, POPs have an organellar targeting peptide at the N-terminus.

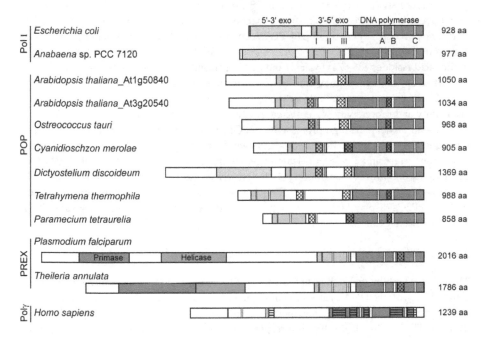

Figure 1. Schematic comparison of the structure of family A DNA polymerases. The colored boxes indicate domains estimated from the Pfam database: pink, 5'-3' exonuclease domain; blue, 3'-5' exonuclease domain; orange, DNA polymerase domain; purple, primase domain; green, helicase domain. Yellow boxes indicate characteristic motifs in the 3'-5' exonuclease and DNA polymerase domains. Thatched boxes represent conserved sequences in POPs. Dotted and striped boxes indicate conserved sequences in PREX and Poly, respectively. In Poly of *Homo sapiens*, a 3'-5' exonuclease domain was not found by Pfam, although 3'-5' exonuclease activity was reported for Poly [6]. This figure was modified from [32] with permission of the publisher.

2.1. Properties of DNA polymerase activity

The properties of DNA polymerase activity of POPs have been examined using recombinant [27,28,31] or native proteins purified from *Cyanidioschyzon* and *Tetrahymena* cells [32,33]. The optimal concentrations of KCl and $MgCl_2$ for DNA polymerase activity are 50-150 and 2.5-5 mM, respectively, which roughly coincide with the values reported in previous studies for organellar DNA polymerases in plants (Table 1). POPs display the highest activity with Poly(dA)/oligo(dT) as a template. Poly(rA)/oligo(dT) could also serve as a template, indicating that POPs have reverse transcriptase activity. Polγ also exhibits reverse transcriptase activity, although the physiological importance of this activity remains to be elucidated.

2.2. Processivity

Processivity is defined as the number of nucleotides added by a DNA polymerase per one binding with the template DNA. POPs, in general, have high processivity values; for example, the processivity of rice recombinant GST-POP and *Cyanidioschyzon* POP is 600-900 nt and 1,300 nt, respectively [28,32]. In comparison, the Klenow fragment of *E. coli* PolI has mid-range processivity of <15 nt [28]. POPs contain three additional internal sequences relative to other family A DNA polymerases (Figure 8). The role of the two extra sequences, amino acid residues 635-674 (Insert I) and 827-852 (Insert II) positioned before motif A (Figure 8-1) and between motif A and motif B (Figure 8-2), were examined in rice POP [28]. DNA binding was decreased in Insert I and II deletion-mutant proteins, while DNA synthesis activity and processivity were decreased only in the POP protein lacking Insert I. These findings suggest that the high processivity of POPs may be due to the existence of the inserted sequences. In animals, $Pol\gamma$ consists of two subunits, a large subunit ($Pol\gamma A$) having DNA polymerase and $3'$-$5'$ exonuclease activities and a small subunit ($Pol\gamma B$) that enhances processivity and primer recognition [34]. Processivity of the *Drosophila* $Pol\gamma A$ subunit is <40 nt, whereas that of $Pol\gamma$ holoenzyme ($Pol\gamma A$ and $Pol\gamma B$) is >1,000 nt [35]. In contrast to animal $Pol\gamma$, POPs display high processivity as a single subunit, and no accessory subunits of POP have been identified to date [28,32].

2.3. Sensitivity to inhibitors

The effects of inhibitors, such as aphidicolin, NEM, dideoxyTTP (ddTTP), and phosphonoacetate (PAA), on the DNA synthesis activity of POPs were evaluated [27,31-33]. Aphidicolin is a specific inhibitor of DNA polymerases α, δ, and ϵ and acts through competition with dCTP or dTTP [36,37]. The sulfhydryl reagent NEM inhibits DNA polymerases α, γ, δ, and ϵ [38], and has a half maximal inhibitory concentration (IC_{50}) of <0.1 mM for $Pol\gamma$. PAA is an analog of pyrophosphate and interacts with viral DNA polymerases and reverse transcriptases at pyrophosphate binding sites to create an alternative reaction pathway [39,40]. ddTTP severely inhibits DNA polymerases β and γ, but only weakly impairs the activities of DNA polymerases δ and ϵ [41]. POPs are not inhibited by aphidicolin or NEM. The inhibitory effect of ddTTP differs depending on the organism, with the IC_{50} ranging from 4-615 μM for POPs (Figure 2A). The activity of POPs is severely inhibited by PAA, as demonstrated by IC_{50} values of 1-25 μM for several POPs (Figure 2B, C). In contrast, other family A DNA polymerases, including PolI and $Pol\gamma$, are not markedly inhibited by PAA, suggesting that PAA is a useful marker for the classification of organellar DNA polymerases in unsequenced eukaryotes. T4 DNA polymerase and DNA polymerase δ of *Saccharomyces cerevisiae*, which are both family B DNA polymerases, are also not sensitive to PAA, but the respective Motif A mutants of each protein, L412M (T4 DNA polymerase) and L612M (DNA polymerase δ of *S. cerevisiae*), are inhibited by PAA [42,43]. The mechanism of inhibition by PAA has not been studied in detail for family A DNA polymerases, and the critical amino acid residues involved in sensitivity to PAA in POPs are unknown due to the limited similarity of family A and B DNA polymerases in the Motif A region.

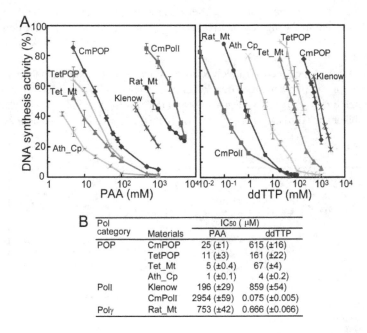

Figure 2. Effect of inhibitors, phosphonoacetic acid (PAA), and dideoxy TTP (ddTTP) on DNA synthesis activity (A). Half maximal inhibitory concentration (IC$_{50}$) for PAA or ddTTP (B). Tet_Mt, *Tetrahymena thermophila* mitochondria; Ath_Cp, *Arabidopsis thaliana* chloroplasts, Klenow, *Escherichia coli* PolI Klenow fragment; CmPolI, *Cyanidioschyzon merolae* PolII; Rat_Mt, rat liver mitochondria. Reproduced from [33] with permission.

2.4. 3'-5' Exonuclease activity

POPs have a 3'-5' exonuclease domain containing three conserved regions, Exo I, Exo II, and Exo III (Figure 1), and this exonuclease activity has been demonstrated in rice [28] and *Cyanidioschyzon* [32]. In rice POP, replacement of Asp365 with Ala in the Exo II domain abolishes nuclease activity, but has no effect on DNA polymerase activity. With regard to 3'-5' exonuclease proofreading activity, POP shows relatively high fidelity for base substitutions (10^{-4} to 10^{-5}; [28]). The primary structure of Polγ appears to lack a 3'-5' exonuclease domain, as indicated by the low E-value of 0.17 for this domain in human Polγ determined using the motif search software Pfam (http://pfam.sanger.ac.uk/). However, Polγ possesses Exo I, Exo II, and Exo III motifs in the N-terminus (Figure 1), and exhibits 3'-5' exonuclease activity and high replication fidelity [6].

2.5. Subcellular localization

POP was first isolated as a plastidial DNA polymerase in rice, and its localization was confirmed by immunoblot analysis using isolated plastids [27]. Subsequent studies using GFP-fusion proteins and/or immunoblotting with isolated plastids and mitochondria demonstrated that POPs are localized to both plastids and mitochondria in *Arabidopsis* and tobacco [31,44], and in the alga *Cyanidioschyzon* [32]. The mitochondrial localization of POP in the ciliate *Tetrahymena* was also determined by immunoblotting [33]. Figure 3 shows all of the known DNA polymerases found in the model plant *A. thaliana* and in humans. The nuclear-localized DNA polymerases involved in genome replication, DNA polymerase α, δ, and ε, are conserved in bikonts and opisthokonts, whereas the nuclear polymerases related to DNA repair differ between organisms. POP and Polγ are the sole replicational DNA polymerases in bikont or opisthokont organelles, where they also act as DNA repair enzymes.

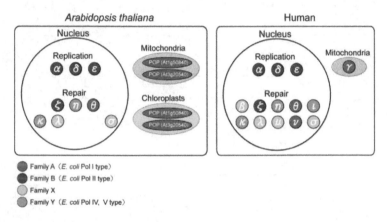

Figure 3. DNA polymerases of a model plant and human. Greek letters in colored circles corresponding to families indicate eukaryotic DNA polymerases alpha to sigma.

2.6. The role of POP *in vivo*

POPs exhibit high processivity and 3'-5' exonuclease activity, and were originally thought to function as organellar DNA replicases. This speculation was verified by analyzing *POP* mutant of *Arabidopsis* [30], whose genome encodes two *POP* genes, *At1g50840* and *At3g20540*, whose protein products are each localized to both plastids and mitochondria (Figure 3). The *At1g50840-At3g20540* double mutant was lethal, while each single mutant had a phenotype characterized as reduced DNA levels in plastids and mitochondria. In addition, only the *At3g20540* mutant displayed elevated sensitivity to ciprofloxacin, which is an inducer of DNA double-strand breaks (DSB). Together, these results show that two distinct POPs are involved in genome replication for plastids and mitochondria, and that the product of *At3g20540* also functions as a DNA repair enzyme in both organelles. In rice, the repair activity of POP was examined by a base excision repair (BER) assay using a recombinant protein, revealing that POP has 5'-deoxyribose phosphate (dRP) lyase activity [28]. Polγ also displays this repair activity [45].

3. Role of POPs in cell-cycle regulation

3.1. Organellar genome replication in plant tissues

Nuclear genomes are replicated during the DNA synthesis phase (S phase), with the daughter genomes being distributed at the mitotic phase (M phase) to maintain ploidy levels. Observations of mitochondrial DNA stained with 4′,6-diamidino-2-phenylindole (DAPI) and microautoradiography using [^{3}H]thymidine have demonstrated that the DNA content and synthesis activity in mitochondria change dramatically during cell proliferation. In the root apical meristem of geranium (*Pelargonium zonale*), mitochondrial DNA in the promeristem, which is located just above the quiescent center, maintain high levels of DNA. However, in the upper root region, located immediately below the elongation zone, mitochondria contain small amounts of DNA [46]. Similar results were reported for the root apical meristem of *Arabidopsis* [47], tobacco [48], and rice [49], shoot apical meristem of *Arabidopsis* [50], and cultured tobacco cells [48,51], in addition to plastids. In *Avena sativa*, plastid DNA is extensively replicated in small cells of shoot apical meristem. Subsequently, as the cells increase in size, plastid numbers increase, while the DNA levels within plastids decrease [52-54]. These results suggest that organellar DNA is predominantly replicated in the meristem, and that the subsequent partition of organellar DNA to daughter cells does not coincide with the synthesis of organellar DNA in cells outside of the meristem center. In multicellular plants, therefore, the replication of organelle genome is not synchronized with the cell cycle or even organellar division.

3.2. Expression of POP in plants

The spatial expression patterns of POPs were analyzed in *Arabidopsis* and rice by *in situ* hybridization, which revealed that *POP* genes are strongly expressed in the apical meristem of roots and shoots, leading to high POP protein levels in these tissues [27,29]. In cultivated tobacco BY-2 cells, the amount of POP transcripts and proteins increases at the initiation of plastidial and mitochondrial DNA replication [31]. These results indicate that POPs function as the organellar genome replicase.

3.3. Red algal cell cycle

The unicellular red alga *C. merolae* contains a single plastid and mitochondrion [55], which both have division cycles that are synchronous with the cell cycle. Synchronous cultures of *Cyanidioschyzon* have been obtained by light-dark cycles [56]. Our group has also performed synchronous culture of *C. merolae* [32,57] using an initial long dark period (30 h) to force the cells into the G1 phase (Figure 4), followed by a 6-h light/18-h dark regime with bubbling with ordinary air. However, due to the low nutrient levels, the conditions were not sufficient to drive the cell cycle (Figure 4A). Two subsequent cycles of 6-h light/18-h dark with a supply of 1% CO_2 enabled the cells to accumulate enough photosynthetic products to allow progression of the cell division cycle, resulting in the synchronous division of cells 4-5 h after the start of the dark period (Figure 4B, C). Therefore, this culture method can discriminate the effects of light from those of the cell cycle in photosynthetic eukaryotes, and contains the cycle in which cellular nutrient level transitions from low to high (Figure 4B).

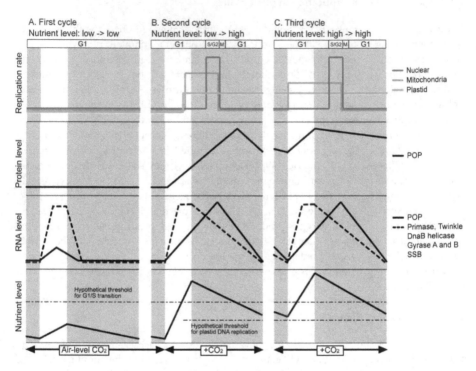

Figure 4. Cell cycle of *Cyanidioschyzon* and the expression of protein or mRNA related to organellar DNA replication. Three cell cycle patterns (A, B, and C) with respect to the nutrient level are shown. The nutrient level was controlled by aeration with or without the addition of CO_2. These drawings are based on the data taken from [57]. The shaded and white areas indicate dark and light cycles, respectively.

We have also determined the replication phases of nuclear, plastid, and mitochondrial DNA by quantitative PCR using cyanobacterial DNA as an internal standard to estimate the absolute amount of DNA (Figure 4, [57]). In the first cell cycle pattern, the level of nuclear and organellar DNA was unaltered (Figure 4A). Nuclear DNA replicated at or near the M-phase in the second and third cycles (Figure 4B, C). The replication of the mitochondrial genome was synchronized with the cell cycle to some extent, with mitochondrial DNA beginning to increase from the middle (second cycle) or beginning (third cycle) of the light phase, and doubling at or near the M-phase, as was observed for nuclear DNA (Figure 4B, C). In contrast, plastid DNA replication continued throughout the entire cell cycle, even after cell division was complete (Figure 4B, C). These results suggest that the replication of nuclear and organellar DNA is initiated after the accumulation of sufficient nutrients by photosynthesis, and that light alone does not serve as a replication signal for nuclear or organellar genomes.

Therefore, *C. merolae* cells may have two checkpoints (or thresholds) based on their nutritional state. The first checkpoint occurs during the G1/S-phase transition. Once cells overcome this point, the nuclear and organellar genomes are targeted for replication. The second threshold is specific for plastid DNA replication. After passage of the first checkpoint for G1/S transition, plastid DNA replication proceeds if the cellular nutrient level exceeds the nutritional threshold required for the replication process.

3.4. Expression of POP in the red algal cell cycle

We determined the expression of POP in synchronous culture of *C. merolae*. The protein level of POP was very low in the first non-dividing cycle (Figure 4A), but continued to increase from the second light period, and subsequently decreased during the dark period (Figure 4B). In the third cycle, the protein level of POP appeared constitutive during the cell cycle, although slight increases in the light phase and decreases in the dark phase were observed (Figure 4C). A small peak in the *POP* mRNA level was detected during the first light period (Figure 4A), with larger peaks appearing soon after entering the dark cycle (Figure 4B, C). The large peaks of *POP* mRNA levels correlated with the rise in mitotic indices.

The transcript level of other possible genes related to organellar DNA replication in *C. merolae* was also examined (Figure 4). Gyrase A and B, which are types of bacterial topoisomerase II, are related to both plastid and mitochondria genome replication in *C. merolae* [58] and *A. thaliana* [59]. SSB is a bacterial single-stranded DNA binding protein that is localized to mitochondria in *A. thaliana* [60]. In plants, DNA primases have not yet been isolated, although primase activity was detected in the chloroplasts of pea and the green alga *Chlamydomonas reinhardtii* [61,62]. DnaB is a bacterial replicational helicase that is encoded in the plastid genome of *C. merolae*. Twinkle is a replicational helicase and is localized to mitochondria in animals. Animal twinkle has only helicase activity; however, it is predicted that twinkle in plants and protists might have both helicase and primase activities [63]. Changes in the expression of these genes were qualitatively similar with each other, and were mainly stimulated by light. The expression pattern of these genes was also similar to that of genes related to photosynthesis, respiration, nuclear DNA repair, and ubiquitin in *C. merolae* [57]. In contrast, the expression pattern of POP transcripts was similar to that of cell cycle regulatory genes, including nuclear replicational DNA polymerase, mitotic cyclin, and mitotic cyclin-dependent kinase (CDK). Based on these findings, it appears that the replication of organellar genomes might be controlled by the expression of POP rather than that of other proteins related to organellar genome replication. Notably, the kinetics of replication differed for plastid and mitochondrial genomes; however, the regulatory mechanisms controlling the replication of the two organelles remain to be elucidated.

4. Possible evolutionary history of organellar DNA polymerases in eukaryotes

POP belongs to family A DNA polymerases, consisting of polymerases harboring sequence similarity to bacterial PolI, such as Polγ, DNA polymerase θ (Polθ), DNA polymerase ν (Polν), and PREX (plastid replication and repair enzyme complex, [64]). Polθ and Polν are DNA repair enzymes and are localized to the nucleus [65,66]. PREX is an apicoplast (plastid like organelle)-localized DNA polymerase in the malaria parasite *Plasmodium falciparum* and contains a DNA polymerase domain, as well as helicase and primase domains (Figure 1 and [67]). Figure 8 shows the alignment of the DNA polymerase domain of several family A DNA polymerases. Although bacterial PolI, POP, and PREX share some homology, POP and PREX contain specific sequences, and the domain structure is clearly different in each polymerase (Figure 1 and Figure 8). Polγ shows low similarity to other family A DNA polymerases, and has many Polγ specific sequences.

Figure 5 shows a phylogenetic tree of family A DNA polymerases. From the tree, it is clear that POPs belong to a well-defined clade that is evolutionarily separated from bacterial PolI. Therefore, it can be concluded that POPs did not originate from PolI of cyanobacteria nor α-proteobacteria. Although PREX may have originated from a red algal secondary endosymbiont, their origin remains unclear, because PREX do not contain POP-specific sequences (Figure 8). POPs are widely conserved in eukaryotes, including amoebozoa, that have a close relationship with opisthokonts in phylogenetic analyses, but POPs have not been detected in opisthokonts, including animals and fungi (Figure 6). This suggests that POP might have originated before the diversification of photosynthetic eukaryotes. Pathogenic protists of animals, including *Blastocystis hominis* and *Perkinsus marinus*, possess POP, while genome-unsequenced pathogens, such as the green alga *Prototheca*, are likely to have POP. Therefore, POP is expected to be a suitable target for killing these pathogens.

From the phylogenetic tree, we proposed an evolutionary model of organellar DNA polymerases (Figure 7). Initially, when the ancestor of eukaryotes acquired mitochondria, the elementary mitochondrial replicase was likely bacterial DNA polymerase III (PolIII) (1 in Figure 7A). PolIII was then replaced by a POP, and the host cell then used POP for the replication of organellar genomes (2 in Figure 7A). We presume that PolIII must have been introduced upon the endosymbiosis event, but another possibility is that an endosymbiont or a host cell had already possessed POP before endosymbiosis. But this idea is considered unlikely because no bacteria having POP have been found so far. In this respect, it is of interest to note that, based on phylogenetic analysis in family A DNA polymerases, it has been postulated that Polγ is of phage origin [69]. POP could also have been acquired from a virus. In effect, the ultimate origin of the ancestral POP is still unknown. The phylogenetic tree (Fig. 5) suggests that the closest relative of POP is Polν or Polθ, which are present in various eukaryotes. It is not impossible then that an ancestral polymerase in eukaryotic host diverged into POP, Polν and Polθ.

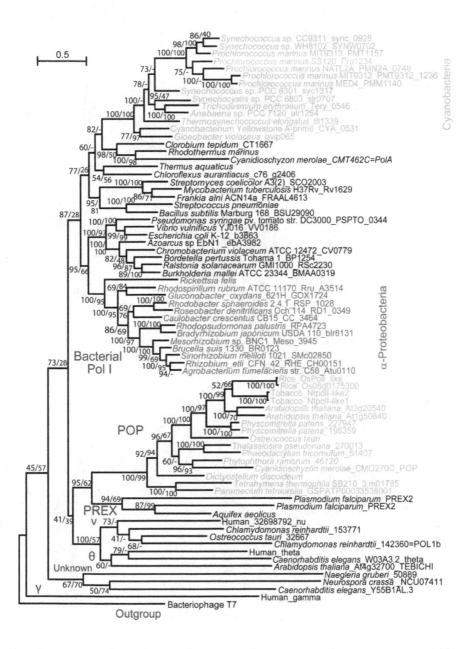

Figure 5. Phylogenetic tree of POPs and other family A DNA polymerases. Reproduced from [32] with permission.

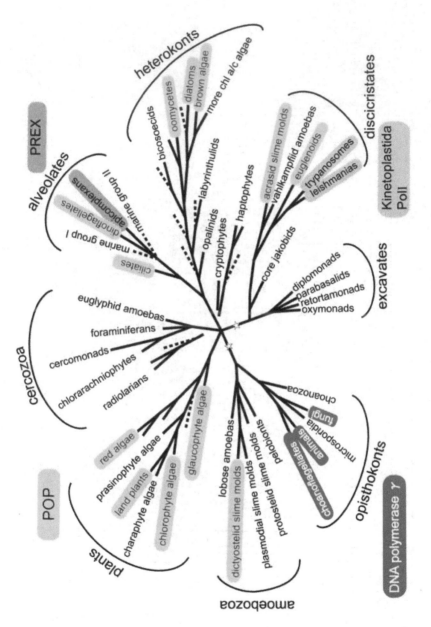

Figure 6. Distribution of organellar DNA polymerases in eukaryotes. Taxons containing POP, Polγ, PREX, and kineto-plastida PolI are enclosed in light green, blue, orange, and purple boxes, respectively. The tree topology in this figure was adapted from [68], and the figure was modified from [33] with permission.

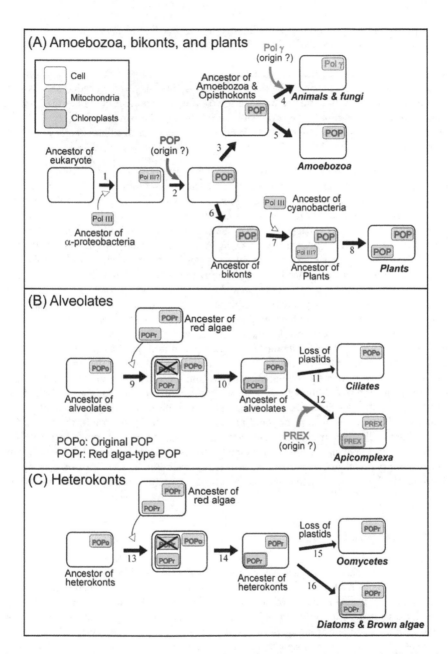

Figure 7. Schematic illustration of the evolution of organellar replication enzymes. The figure was modified from [33] with permission.

In the plastids of plants and algae, POP also replaced PolIII, and thus POPs are presently found in most eukaryotes (3-5 and 6-8 in Figure 7A). In opisthokonts, however, POP was replaced by Polγ, whose origin is also unknown (4 in Figure 7A). Chromalveolates, consisting of alveolates and heterokonts such as diatoms, must have had a POP for mitochondrial replication before the occurrence of secondary endosymbiosis. Phylogenetic analysis suggests that the POPs of diatoms are more closely related to red algal POP than the POPs of ciliate *Tetrahymena* (Figure 5). The original POP might have been replaced by the POP of a red algal endosymbiont in diatoms (13-16 in Figure 7C), whereas in ciliates, the original POP has been retained (9-11 in Figure 7B).

Based on the genomic data obtained to date, Polγ is found only in opisthokonts, indicating that two different polymerases cannot co-exist, at least over a long evolutionary span. The catalytic subunit of bacterial PolIII is also not encoded by eukaryotic genomes, although the PolIII gamma subunit, which functions as a clamp loader in bacteria, is conserved in land plants, such as *A. thaliana*, which has three gamma subunits, At1g14460, At2g02480, and At4g24790 [13]. One of the possible reasons why PolIII was replaced by POP may be the fact that POP is a single polypeptide enzyme, whereas PolIII consists of ten subunits. Therefore, the nuclear control of organellar DNA replication would be easier with nuclear-encoded POP. This also raises the question: why was POP replaced with Polγ? Unfortunately, although we do not have a clear answer for this question, the replacement event might be related to the mechanism of organellar genome replication. In animals, three replication modes have been proposed: the classical strand-displacement replication mode, a strand-coupled mode, and a RITOLS (ribonucleotide incorporation throughout the lagging strand) mode [70]. This contrasts with plant plastids, for which at least two modes of replication have been proposed, namely rolling circle replication via a D-loop and recombination-dependent replication [71]. Although the proposed replication modes in animals and plants remain to be confirmed, it is likely that the type of replication mode is different in the organelles of animals (opisthokonts) and plants (bikonts). In opisthokonts, the replication mode of organellar genomes of animals may have arisen before the replacement of POP with Polγ, with Polγ being a suitable enzyme for the replication process of animals. Secondary or tertiary endosymbionts do not exist among opisthokonts, a fact that may be due to differences in the organellar genome replication mode or organellar DNA polymerase type.

5. Conclusion and prospects

POPs have been isolated as organellar-specific DNA polymerases in a number of photosynthetic eukaryotes and ciliates. As the majority of biologists still believe that all mitochondrial replication enzymes are Polγ, the primary objective of this review was to introduce POP to the wider research community. Although both POP and Polγ are family A DNA polymerases, their primary structures are quite different from one another. However, POP and Polγ display similar DNA polymerase activities that are characteristics of replicases, including high processivity, 3'-5' exonuclease activity, and reverse transcriptase activity. Eukaryotes containing a POP gene do not have a gene for Polγ, and *vice versa*. In our hypothesis concerning the transitional evolution of organellar DNA polymerase in eukaryotes, POP was proposed to be the primary organellar replicase and was then replaced by Polγ in opisthokonts. POP might have been replaced by PREX and kinetoplastida PolI in apicomplexa and trypanosomes, respectively. Phylogenetic evidence suggests that organellar DNA polymerases are easily replaced, unlike nuclear replicational DNA polymerases, which are conserved in all eukaryotes.

The sensitivity of POP to DNA polymerase inhibitors clearly differs from that of Polγ. To date, POPs have been shown to be commonly inhibited by phosphonoacetate. The inhibition mechanisms remained unclear for family A DNA polymerases, including POP, although it was reported that motif A in the polymerase domain of family B DNA polymerases is involved in the sensitivity to phosphonoacetate [42,43]. The detailed study of the inhibitory mechanisms and structural analysis of POP are needed, although POP is likely to be conserved in pathogenic bikonts, such as the green alga *Prototheca* and chromalveolata *Blastocystis*. Determining the structural differences in essential enzymes between a pathogen and host, and identifying pathogen-specific enzymes with no homologues in a host may identify suitable targets for chemotherapy. Such an approach is needed for targeting the malaria parasite. Chloroquine, mefloquine, and quinine have been used as antimalarial drugs. These reagents inhibit the production of the malarial pigment hemozoin. In addition, dihydrofolate reductase (DHFR) of malaria parasite is inhibited by proguanil and pyrimethamine. However, drug-resistant mutants of the parasite have emerged, and a new drug and enzyme target are therefore needed [72]. An apicoplast is non-photosynthetic plastid-like organelle that contains 27-35 kb of DNA in apicomplexa, and DNA replication within apicoplasts may be a good drug target, because apicoplasts harbors various essential metabolic pathways, such as those involving fatty acids, isoprenoid, and heme [73]. In plants and protists, our knowledge of the supporting players of organellar DNA replication, such as primase, helicase, topoisomerase, and single-stranded DNA binding protein (SSB), are limited. To understand the mechanism and regulation of replication in plastids and mitochondria, it is necessary that the composition of these enzymes in each organelle be determined. In addition, reconstitution of the replicational machinery of each organellar genome should be attempted. In humans, successful *in vitro* reconstitution of the mitochondrial DNA replisome, including Polγ, twinkle helicase, and SSB, was demonstrated [74]. The further development of organellar replisome models in plants and protists may pave the way for greater understanding of the replication mode and discovery of new antiprotozoan reagents.

In multicellular plants, genomes of organelles are replicated in meristematic tissues, but the process is not synchronous with the cell cycle or even with organellar division. In the unicellular red alga *C. merolae*, which contains a single plastid and mitochondrion, the expression of POP appears constitutive during the cell cycle. POP is localized in both organelles, but the kinetics of replication differs for plastid and mitochondrial genomes. Replication of the mitochondrial genome is synchronous with the cell cycle to a certain extent, whereas replication of the plastid genome continues throughout the entire cell cycle. The organellar replication is regulated by cellular nutrient levels, and POP protein levels are closely correlated with nutrient levels. The mechanisms regulating the replication of plastids and mitochondria represent a new and exciting area of research in cell biology.

Appendix

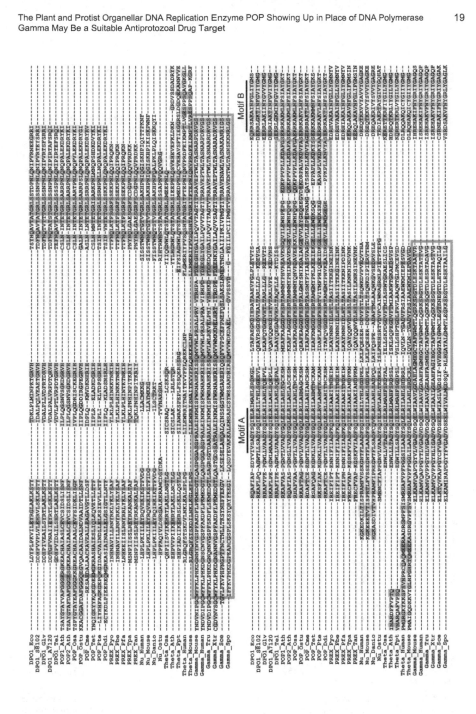

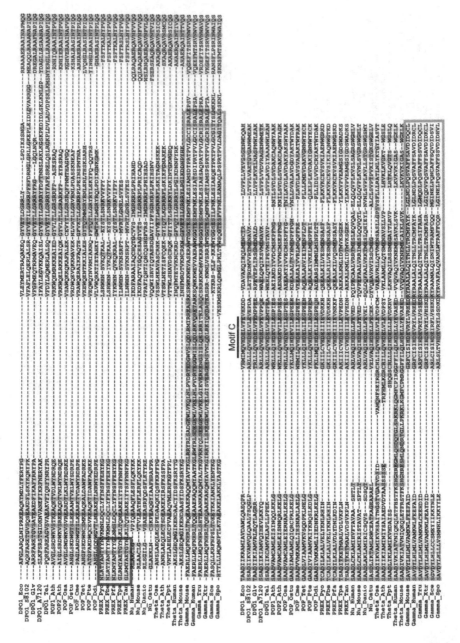

Figure 8. Alignment of the DNA polymerase domain of family A DNA polymerases. Green, blue, and orange boxes show specific sequences of POPs, DNA polymerase nu, and DNA polymerase gamma, respectively. Eco, *Escherichia coli*;

S8102, *Synechococcus* sp. WH8102; Glv, *Gloeobacter violaceus*; A7120, *Anabaena* sp. PCC 7120; Tel, *Thermosynechococcus elongates*; Ath, *Arabidopsis thaliana*; Osa, *Oryza sativa*; Ostu, *Ostreococcus tauri*, Cme, *Cyanidioschyzon merolae*; Tet, *Tetrahymena thermophila* SB210; Pte, *Paramecium tetraurelia*; Ddi, *Dictyostelium discoideum*; Pyo, *Plasmodium yoelii*; Pbe, *Plasmodium berghei*; Pfa, *Plasmodium falciparum*; Tpa, *Theileria parva*; Tan, *Theileria annulata*; Danio, *Danio rerio*; Ppt, *Physcomitrella patens*; Tru, *Takifugu rubripes*; Xla, *Xenopus tropicalis*; Sce, *Saccharomyces cerevisiae*; Spo, *Schyzosaccharomyces pombe*. Reproduced with permission [33].

Acknowledgements

This work was supported in part by Grants from Core Research for Evolutional Science and Technology (CREST) from the Japan Science and Technology Agency (JST), Japan, the Global Center of Excellence (GCOE) Program "From the Earth to 'Earths'" from the MEXT, Japan, and the Canon Foundation.

Author details

Takashi Moriyama[1,2*] and Naoki Sato[1,2*]

*Address all correspondence to:

*Address all correspondence to: naokisat@bio.c.u-tokyo.ac.jp

1 Department of Life Sciences, Graduate School of Arts and Sciences, The University of Tokyo, Tokyo, Japan

2 JST, CREST, Gobancho, Chiyoda-ku, Tokyo, Japan

References

[1] Margulis, L., & Bermudes, D. (1985). Symbiosis as a mechanism of evolution: status of cell symbiosis theory. *Symbiosis*, 1, 101-124.

[2] Douglas, S. E. (1998). Plastid evolution: Origins, diversity, trends. *Curr Opin Genet Dev*, 8(6), 655-661.

[3] Gray, M. W. (1999). Evolution of organellar genomes. *Curr Opin Genet Dev*, 9(6), 678-687.

[4] Wang, Y., Farr, C. L., & Kagni, L. S. (1997). Accessory subunit of mitochondrial DNA polymerase from Drosophila embryos. Cloning, molecular analysis, and association in the native enzyme. *J Biol Chem*, 272(21), 13640-13646.

[5] Lecrenier, N., & Foury, F. (2000). New features of mitochondrial DNA replication system in yeast and man. *Gene*, 246(1-2), 37-48.

[6]   Kaguni, L. S. (2004). DNA polymerase gamma, the mitochondrial replicase. *Annu Rev Biochem*, 73, 293-320.

[7]   Tewari, K. K., & Wildman, S. G. (1967). DNA polymerase in isolated tobacco chloroplasts and nature of the polymerized product. *Proc Natl Acad Sci U S A*, 58(2), 689-696.

[8]   Spencer, D., & Whitfeld, P. R. (1967). DNA synthesis in isolated chloroplasts. *Biochem Biophys Res Commun*, 28(4), 538-542.

[9]   Wintersberger, E. (1966). Occurrence of a DNA-polymerase in isolated yeast mitochondria. *Biochem Biophys Res Commun*, 25(1), 1-7.

[10] Parsons, P., & Simpson, M. V. (1967). Biosynthesis of DNA by isolated mitochondria: incorporation of thymidine triphosphate-2-C-14. *Science*, 155(3758), 91-93.

[11] Sakai, A., Suzuki, T., Nagata, N., Sasaki, N., Miyazawa, Y., Saito, C., Inada, N., Nishimura, Y., & Kuroiwa, T. (1999). Comparative analysis of DNA synthesis activity in plastid-nuclei and mitochondrial-nuclei simultaneously isolated from cultured tobacco cells. *Plant Sci*, 140(1), 9-19.

[12] Sakai, A. (2001). In vitro transcription/DNA synthesis using isolated organelle-nuclei: Application to the analysis of the mechanisms that regulate organelle genome function. *J Plant Res*, 114(2), 199-211.

[13] Sakai, A., Takano, H., & Kuroiwa, T. (2004). Organelle nuclei in higher plants: structure, composition, function, and evolution. *Int Rev Cytol*, 238, 59-118.

[14] Keller, S. J., Biedenbach, S. A., & Meyer, R. R. (1973). Partial purification of a chloroplast DNA polymerase from Euglena gracilis. *Biochem Biophys Res Commun*, 50(3), 620-628.

[15] Castroviejo, M., Tharaud, D., Tarrago-Litrak, L., & Litvak, S. (1979). Multiple deoxyribonucleic acid polymerases from quiescent wheat embryos. Purification and characterization of three enzymes from the soluble cytoplasm and one from purified mitochondria. *Biochem J*, 181(1), 183-191.

[16] Fukasawa, H., & Chou, M. Y. (1980). Mitochondrial DNA polymerase from Cauliflower inflorescence. *Jpn J Genet*, 55(6), 441-445.

[17] Sala, F., Amileni, A. R., Parisi, B., & Spadari, S. (1980). A gamma-like DNA polymerase in spinach chloroplasts. *Eur J Biochem*, 112(2), 211-217.

[18] Christophe, L., Tarrago-Litvak, L., Castroviejo, M., & Litvak, S. (1981). Mitochondrial DNA polymerase from wheat embryos. *Plant Sci Lett*, 21(2), 181-192.

[19] McKown, R. L., & Tewari, K. K. (1984). Purification and properties of a pea chloroplast DNA polymerase. *Proc Natl Acad Sci U S A*, 81(8), 2354-2358.

[20] Heinhorst, S., Cannon, G. C., & Weissbach, A. (1990). Chloroplast and mitochondrial DNA polymerases from cultured soybean cells. *Plant physiol*, 92(4), 939-945.

[21] Keim, C. A., & Mosbaugh, D. W. (1991). Identification and characterization of a 3' to 5' exonuclease associated with spinach chloroplast DNA polymerase. *Biochemistry*, 30(46), 11109-11118.

[22] Wang, Z. F., Yang, J., Nie, Z. Q., & Wu, M. (1991). Purification and characterization of a gamma-like DNA polymerase from Chlamydomonas reinhardtii. *Biochemistry*, 30(4), 1127-1131.

[23] Meißner, K., Heinhorst, S., Cannon, G. C., & Börner, T. (1993). Purification and characterization of a gamma-like DNA polymerase from Chenopodium album L. *Nucleic Acids Res.*, 21(21), 4893-4899.

[24] Bailey, J. C., Heinhorst, S., & Cannon, G. C. (1995). Accuracy of deoxynucleotide incorporation by soybean chloroplast DNA polymerases is independent of the presence of a 3' to 5' exonuclease. *Plant Physiol*, 107(4), 1277-1284.

[25] Gaikwad, A., Hop, D. V., & Mukherjee, S. K. (2002). A 70-kDa chloroplast DNA polymerase from pea (Pisum sativum) that shows high processivity and displays moderate fidelity. *Mol Genet Genomics*, 267(1), 45-56.

[26] Moriyama, T. (2008). Studies on the DNA polymerases localized to plastids and mitochondria in plants and algae. Thesis. The University of Tokyo;.

[27] Kimura, S., Uchiyama, Y., Kasai, N., Namekawa, S., Saotome, A., Ueda, T., Ando, T., Ishibashi, T., Oshige, M., Furukawa, T., Yamamoto, T., Hashimoto, J., & Sakaguchi, K. (2002). A novel DNA polymerase homologous to Escherichia coli DNA polymerase I from a higher plant, rice (Oryza sativa L.). *Nucleic Acids Res*, 30(7), 1585-1592.

[28] Takeuchi, R., Kimura, S., Saotome, A., & Sakaguchi, K. (2007). Biochemical properties of a plastidial DNA polymerase of rice. *Plant Mol Biol*, 64(5), 601-611.

[29] Mori, Y., Kimura, S., Saotome, A., Kasai, N., Sakaguchi, N., Uchiyama, Y., Ishibashi, T., Yamamoto, T., Chiku, H., & Sakaguchi, K. (2005). Plastid DNA polymerases from higher plants, Arabidopsis thaliana. *Biochem. Biophys. Res Commun*, 334(1), 43-50.

[30] Parent, J. S., Lepage, E., & Brisson, N. (2011). Divergent roles for the two PolI-like organelle DNA polymerases of Arabidopsis. *Plant Physiol*, 156(1), 254-262.

[31] Ono, Y., Sakai, A., Takechi, K., Takio, S., Takusagawa, M., & Takano, H. (2007). NtPoII-like1 and NtPoII-like2, bacterial DNA polymerase I homologues isolated from BY-2 cultured tobacco cells, encode DNA polymerases engaged in DNA replication in both plastids and mitochondria. *Plant Cell Physiol*, 48(12), 1679-1692.

[32] Moriyama, T., Terasawa, K., Fujiwara, M., & Sato, N. (2008). Purification and characterization of organellar DNA polymerases in the red alga Cyanidioschyzon merolae. *FEBS J*, 275(11), 2899-2918.

[33] Moriyama, T., Terasawa, K., & Sato, N. (2011). Conservation of POPs, the plant organellar DNA polymerases, in eukaryotes. *Protist*, 162(1), 177-187.

[34] Wanrooij, S., & Falkenberg, M. (2010). The human mitochondrial replication fork in health and disease. *Biochim Biophys Acta*, 1797(8), 1378-1388.

[35] Williams, A. J., Wernette, C. M., & Kaguni, L. S. (1993). Processivity of mitochondrial DNA polymerase from *Drosophila* embryos. Effects of reaction conditions and enzyme purity. *J Biol Chem*, 268(33), 24855-24862.

[36] Holmes, A. M. (1981). Studies on the inhibition of highly purified calf thymus 8S and 7.3S DNA polymerase alpha by aphidicolin. *Nucleic Acids Res*, 9(1), 161-168.

[37] Weiser, T., Gassmann, M., Thömmes, P., Ferrari, E., Hafkemeyer, P., & Hübscher, U. (1991). Biochemical and functional comparison of DNA polymerases alpha, delta, and epsilon from calf thymus. *J Biol Chem*, 266(16), 10420-10428.

[38] Chavalitshewinkoon-Petmitr, P., Chawprom, S., Naesens, L., Balzarini, J., & Wilairat, P. (2000). Partial purification and characterization of mitochondrial DNA polymerase from Plasmodium falciparum. *Parasitol Int*, 49(4), 279-288.

[39] Shiraki, K., Okuno, T., Yamanishi, K., & Takahashi, M. (1989). Phosphonoacetic acid inhibits replication of human herpesvirus-6. *Antiviral Res*, 12(5-6), 311-318.

[40] Leinbach, S. S., Reno, J. M., Lee, L. F., Isbell, A. F., & Boezi, J. A. (1976). Mechanism of phosphonoacetate inhibition of herpesvirus-induced DNA polymerase. *Biochemistry*, 15(2), 426-430.

[41] Kornberg, A., & Baker, T. A. (1992). DNA Replication. 2nd ed. New York: W.H. Freeman and Company, 199.

[42] Reha-Krantz, L. J., & Nonay, R. L. (1994). Motif A of bacteriophage T4 DNA polymerase: role in primer extension and DNA replication fidelity. Isolation of new antimutator and mutator DNA polymerases. *J Biol Chem*, 269(8), 5635-5643.

[43] Li, L., Murphy, K. M., Kanevets, U., & Reha-Krantz, L. J. (2005). Sensitivity to phosphonoacetic acid: a new phenotype to probe DNA polymerase δ in Saccharomyces cerevisiae. *Genetics*, 170(2), 569-580.

[44] Christensen, A. C., Lyznik, A., Mohammed, S., Elowsky, C. G., Elo, A., Yule, R., & Mackenzie, S. A. (2005). Dual-domain, dual-targeting organellar protein presequences in Arabidopsis can use non-AUG start codons. *Plant Cell*, 17(10), 2805-2816.

[45] Pinz, K. G., & Bogenhagen, D. F. (2000). Characterization of a catalytically slow AP lyase activity in DNA polymerase γ and other family A DNA polymerases. *J Biol Chem*, 275(17), 12509-12514.

[46] Kuroiwa, T., Fujie, M., & Kuroiwa, H. (1992). Studies on the behavior of mitochondrial DNA: Synthesis of mitochondrial DNA occurs actively in a specific region just above the quiescent center in the root apical meristem of Pelargonium zonale. *J Cell Sci*, 101(3), 483-493.

[47] Fujie, M., Kuroiwa, H., Kawano, S., & Kuroiwa, T. (1993). Studies on the behavior of organelles and their nucleoids in the root apical meristem of Arabidopsis thaliana (L.) Col. *Planta*, 189(3), 443-452.

[48] Suzuki, T., Kawano, S., Sakai, A., Fujie, M., Kuroiwa, H., Nakamura, H., & Kuroiwa, T. (1992). Preferential mitochondrial and plastid DNA synthesis before multiple cell divisions in Nicotiana tabacum. *J Cell Sci*, 103(3), 831-837.

[49] Suzuki, T., Sasaki, N., Sakai, A., Kawano, S., & Kuroiwa, T. (1995). Localization of organelle DNA synthesis within root apical meristem of rice. *J Exp Bot*, 46(1), 19-25.

[50] Fujie, M., Kuroiwa, H., Kawano, S., Mutoh, S., & Kuroiwa, T. (1994). Behavior of organelles and their nucleoids in the shoot apical meristem during leaf developmet in Arabidopsis thaliana L. *Planta*, 194(3), 395-405.

[51] Okamura, S., Suzuki, T., Miyazawa, Y., Kuroiwa, T., & Sakai, A. (2002). Activation of organelle DNA synthesis during the initial phase of proliferation of BY-2 cultured tobacco cells after medium renewal. *Plant Morphol*, 14(1), 16-28.

[52] Hashimoto, H. (1986). Double ring structure around the constricting neck of dividing plastids of Avena sativa. *Protoplasma*, 135(2-3), 166-172.

[53] Possingham, J. V., Hashimoto, H., & Oross, J. (1988). Factors that influence plastid division in higher plants. In Boffey SA, Lloyd D. (ed.), *The division and segregation of organelles.*, New York: W.H. Cambridge University Press.

[54] Hashimoto, H., & Possingham, J. V. (1989). DNA Levels in Dividing and Developing Plastids in Expanding Primary Leaves of Avena sativa. *J Exp Bot*, 40(2), 257-262.

[55] Matsuzaki, M., Misumi, O., Shin-i, T., Maruyama, S., Takahara, M., Miyagishima, S., Mori, T., Nishida, K., Yagisawa, F., Nishida, K., Yoshida, Y., Nishimura, Y., Nakao, S., Kobayashi, T., Momoyama, Y., Higashiyama, T., Minoda, A., Sano, M., Nomoto, H., Oishi, K., Hayashi, H., Ohta, F., Nishizaka, S., Haga, S., Miura, S., Morishita, T., Kabeya, Y., Terasawa, K., Suzuki, Y., Ishii, Y., Asakawa, S., Takano, H., Ohta, N., Kuroiwa, H., Tanaka, K., Shimizu, N., Sugano, S., Sato, N., Nozaki, H., Ogasawara, N., Kohara, Y., & Kuroiwa, T. (2004). Genome sequence of the ultrasmall unicellular red alga Cyanidioschyzon merolae 10D. *Nature*, 428(6983), 653-657.

[56] Suzuki, K., Ehara, T., Osafune, T., Kuroiwa, H., Kawano, S., & Kuroiwa, T. (1994). Behavior of mitochondria, chloroplasts and their nuclei during the mitotic cycle in the ultramicroalga Cyanidioschyzon merolae. *Eur J Cell Biol*, 63(2), 280-288.

[57] Moriyama, T., Terasawa, K., Sekine, K., Toyoshima, M., Koike, M., Fujiwara, M., & Sato, N. (2010). Characterization of cell-cycle-driven and light-driven gene expression in a synchronous culture system in the unicellular rhodophyte Cyanidioschyzon merolae. *Microbiology*, 156(6), 1730-1737.

[58] Itoh, R., Takahashi, H., Toda, K., Kuroiwa, H., & Kuroiwa, T. (1997). DNA gyrase involvement in chloroplast-nucleoid division in Cyanidioschyzon merolae. *Eur J Cell Biol*, 73(3), 252-258.

[59] Wall, M. K., Mitchenall, L. A., & Maxwell, A. (2004). Arabidopsis thaliana DNA gyrase is targeted to chloroplasts and mitochondria. *Proc Natl Acad Sci U S A*, 101(20), 7821-7826.

[60] Edmondson, A. C., Song, D., Alvarez, L. A., Wall, M. K., Almond, D., Mc Clellan, D. A., Maxwell, A., & Nielsen, B. L. (2005). Characterization of a mitochondrially targeted single-stranded DNA-binding protein in Arabidopsis thaliana. *Mol Genet Genomics.*, 273(2), 115-122.

[61] Nielsen, B. L., Rajasekhar, V. K., & Tewari, K. K. (1991). Pea chloroplast DNA primase: characterization and role in initiation of replication. *Plant Mol Biol*, 16(6), 1019-1034.

[62] Nie, Z., & Wu, M. (1999). The functional role of a DNA primase in chloroplast DNA replication in Chlamydomonas reinhardtii. *Arch Biochem Biophys*, 369(1), 174-180.

[63] Shutt, T. E., & Gray, M. W. (2006). Twinkle, the mitochondrial replicative DNA helicase, is widespread in the eukaryotic radiation and may also be the mitochondrial DNA primase in most eukaryotes. *J Mol Evol*, 62(5), 588-599.

[64] Mukhopadhyay, A., Chen, C. Y., Doerig, C., Henriquez, F. L., Roberts, C. W., & Barrett, M. P. (2009). The Toxoplasma gondii plastid replication and repair enzyme complex, PREX. *Parasitology*, 136(7), 747-755.

[65] Seki, M., & Wood, R. D. (2008). DNA polymerase θ (POLQ) can extend from mismatches and from bases opposite a (6-4) photoproduct. *DNA Repair (Amst)*, 7(1), 119-127.

[66] Marini, F., Kim, N., Schuffert, A., & Wood, R. D. (2003). POLN, a nuclear PolA family DNA polymerase homologous to the DNA cross-link sensitivity protein Mus308. *J Biol Chem*, 278(34), 32014-32019.

[67] Seow, F., Sato, S., Janssen, C. S., Riehle, M. O., Mukhopadhyay, A., Phillips, R. S., Wilson, R. J., & Barrett, M. P. (2005). The plastidic DNA replication enzyme complex of Plasmodium falciparum. *Mol Biochem Parasitol*, 141(2), 145-153.

[68] Baldauf, S. L. (2003). The deep roots of eukaryotes. *Science* , 300(5626), 1703-1706.

[69] Filée, J., Forterre, P., Sen-Lin, T., & Laurent, J. (2002). Evolution of DNA polymerase families: evidences for multiple gene exchange between cellular and viral proteins. *J Mol Evol*, 54(6), 763-773.

[70] Wanrooij, S., & Falkenberg, M. (2010). The human mitochondrial replication fork in health and disease. *Biochim Biophys Acta*, 1797(8), 1378-1388.

[71] Nielsen, B. L., & Cupp, Brammer. J. (2010). Mechanisms for maintenance, replication, and repair of the chloroplast genome in plants. *J Exp Bot*, 61(10), 2535-2537.

[72] Guiguemde, W. A., Shelat-Bustos, Garcia., Diagana, J. F., Gamo, T. T., Guy, F. J., & , R. K. (2012). Global phenotypic screening for antimalarials. *Chem Biol*, 19(1), 116-129.

[73] García-Estrada, C., Prada, C. F., Fernández-Rubio, C., Rojo-Vázquez, F., & Balaña-Fouce, R. (2010). DNA topoisomerases in apicomplexan parasites: promising targets for drug discovery. *Proc Biol Sci*, 277(1689), 1777-1787.

[74] Korhonen, J. A., Pham, X. H., Pellegrini, M., & Falkenberg, M. (2004). Reconstitution of a minimal mtDNA replisome in vitro. *EMBO J*, 23(12), 2423-2429.

Chapter 2

Replicational Mutation Gradients, Dipole Moments, Nearest Neighbour Effects and DNA Polymerase Gamma Fidelity in Human Mitochondrial Genomes

Hervé Seligmann

Additional information is available at the end of the chapter

1. Introduction

The large amount of available human mitochondrial genome data from Ingman and Gyllenstein and from Ruiz-Pesini et al [see http://www.mtdb.igp.uu.se/, 1 and http://www.mitomap.org/MITOMAP, 2] enables to study in some detail the spectrum of mutations observed within this species' mitochondrion. DNA mutations have two main causes: spontaneous chemical alterations of nucleotides, from one nucleotide 'species' to another, such as hydrolytic deaminations from C->T and A to hypoxanthine, which pairs with C and leads to its replacement by G (in the following summarized as A->G); and inaccuracies by the enzymatic machinery that is responsible for the polymerization of new DNA strands from the template of the existing DNA during DNA replication. Here I explore the tendency for mutations from different genes and mutation types to be explained by the first (physico-chemical), or the other (more enzymatic/biological) factor, also in relation to adaptive constraints (natural selection is weaker against DNA mutations that cause no or only conservative changes at the protein level). The relative importance of these various factors affecting mutation spectra is investigated for observed human mitochondrial mutations in relation to different types of substitutions and different genes. I also explore nearest neighbour effects on the different mutation types, though the relative contribution of this factor in relation to others is not evaluated here.

The main, presumably sole DNA replicating enzyme in vertebrate mitochondria is the DNA gamma polymerase, which evolved from a bacterial tRNA synthetase [3]. Twelve types of substitutions of one nucleotide by another nucleotide occur at different frequencies. The most frequent changes occur within each of the nucleotide families, purines (adenosine, A, and guanine, G) and pyrimidines (cytosine, C, and thymidine, T), as these involve less changes in molecular structure. These four purine to purine or pyrimidine to pyrimidine mutations are called transitions, the eight mutations from one chemical group to another are called transversions. A simplistic model predicts the frequencies of all substitutions, based on the dipole moment of the nucleotides [4], for a DNA region supposed to have no function, a pseudogene [5], so that observed substitution frequencies are believed unaffected by natural selection. The partial dipole moment of a chemical bond is proportional to the distance of an electron's mean position, in the chemical bond between atoms, from mid-distance between these atoms. The dipole moment of a molecule is the product of all partial dipole moments. G and C have high dipole moments, A and T have low dipole moments [6]. The hypothesized model assumes that a high dipole moment indicates high chemical reactivity, and hence probable alteration by chemical processes. Indeed, observed frequencies of mutations in pseudogenes of one nucleotide into another nucleotide are proportional to dipole moment changes: nucleotides with low dipole moment substitute those with high dipole moment.

Independently of the dipole moment hypothesis, some spontaneous chemical reactions, deaminations, A->G and C->T, occur preferentially while the heavy DNA strand is in the single stranded state [7, 8]. This occurs mainly during replication and transcription (DNA and RNA polymerization). Distances from replication origins and for transcription, from promoters [9, 10], determine durations that different DNA regions remain single stranded, creating gradients in deaminations in genomes with asymmetric replication, such as mitochondrial genomes (reviewed in [11, 12]). Hence gene position affects transition frequencies. Site-specific mutation rates estimated by phylogenetic reconstruction suggest that mutation gradients might also exist for some transversions [13, 14], indicating that single-strandedness might affect also substitutions that are not A->G or C->T.

Here I analyse mutation patterns observed in the 13 human mitochondrial protein coding genes, to estimate relative contributions of different processes to observed mutation patterns: replicational gradients [13, 14, 15], dipole moments [6, 16], selection against mutations that alter coding properties at the protein level [17] and gamma polymerase misincorporation [18], and potential interactions between these processes. I also explore nearest neighbour effects on mutation rates. The present analyses are also original in the sense that they are based on comparative analyses of sequence data, all the data originating from a single species, and not, as previous ones (i.e. [13, 14]), from comparisons between different, frequently evolutionarily distant species.

2. Dipole moments and the accuracy of the DNA gamma polymerase

The estimation of process-specific contributions of different mechanisms to a given phenomenon requires considering dependence among processes. Dipole moment effects are independent of the deamination gradient model which predicts a gradient in C->T, consistent with the model of dipole moment decrease, but an A->G gradient is inconsistent. This means that one deamination gradient fits into the dipole moment model, and the other does not, making both approaches approximately unrelated.

The issue of the accuracy of the gamma polymerase is more complex. It is indeed plausible that nucleotide misinsertion results from misrecognition of nucleotides by the polymerase, the latter due to physico-chemical similarities between nucleotides. This principle is also suggested by the high average misincorporation rates resulting in transitions (447.25±375.22) as compared to misincorporation rates causing transversions (121.64±186.6), explaining 33 percent of the variation in rates between different misincorporation types [18]. If so, the absolute value of the difference between dipole moments of nucleotides (from [16]) should be inversely proportional to misincorporation rates ('kd' in [18]), high rates occurring for nucleotides with similar dipole moments. This model differs from the model of dipole moment decrease, as it deals with the absolute value of the difference between dipole moments, and not the signed difference.

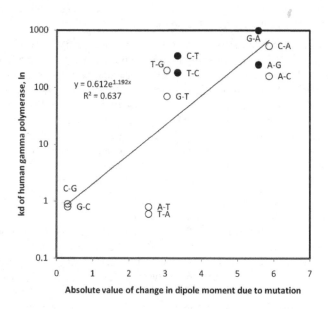

Figure 1. Misincorporation versus absolute difference between dipole moment of substituted and substituting nucleotide. Transitions (filled symbols) have high kds, but similar dipole moments decrease misincorporation kds.

The dipole similarity model for polymerase misincorporation rates can be dismissed at this point. Misincorporation rates increase, not decrease as expected, with absolute values of differences between dipole moments (r = 0.80, Figure 1). This unexplained association could reflect effects of other properties on misincorporation rates, properties that are inversely correlated with dipole moments. Note that after controlling for differences between transitions and transversions, the correlation shown in Figure 1 decreases (r = 0.54, not shown), yet the analysis confirms the principle that nucleotide substitutions with high kds tend to be substitutions between nucleotides with highly divergent dipole moments.

It is also possible that many nucleotide misincorporations result from the delay occurring between nucleotide recognition by the gamma polymerase and its incorporation in the elongating DNA polymer. One could suppose that some misincorporations are not due to misrecognitions, but to spontaneous mutations occurring after the nucleotide's accurate recognition by the polymerase, and before its incorporation. In that case, misincorporation rates should match the dipole moment model for decreased dipole moment: high rates are observed when substitutions decrease the dipole moment. This hypothesis cannot be ruled out, as misincorporation rates increase with the signed difference between nucleotide dipole moments (r = 0.50, not shown). Controlling for differences in kd between transitions and transversions, this positive association increases (r = 0.60, Figure 2).

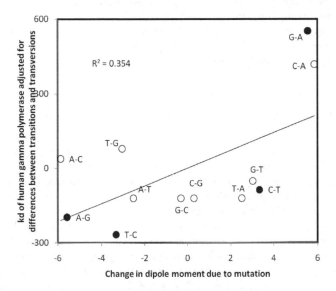

Figure 2. Adjusted misincorporation kd as a function of difference between dipole moments of substituted and substituting nucleotide. High kds imply dipole moment decrease.

Note that if causal interpretations of the associations in Figures 1 and 2 are relevant, it would be the dipole moment that affects kds. An alternative explanation to the trend in Figure 2 is that the gamma polymerase binds more readily nucleotides with low than high dipole moment, hence resulting in this biased misincorporation trend. Such a pattern could easily be caused by an overall relatively hydrophobic nature of the residues that constitute the polymerase's binding site (low dipole moment implying relative hydrophobicity). Even a very small bias for hydrophobic interactions would cause strong biases in analyses focusing on misincorporation rates. However, this hydrophobicity hypothesis does not seem to fit, at least in its simplistic form, what is known about the active site of the gamma polymerase according to the crystal structure published by Lee et al [19]. The active site consists of amino acids E895, Y951, R943 and Y955, among which one residue is positively charged (E, glutamic acid), one negatively (R, arginine) and two are hydrophilic (Y, tyrosine). Note that none is classified as a hydrophobic amino acid. Hence the positive association in Figure 2 does not seem explained by active site hydrophobicity. Speculatively, electrostatic neutrality could favour misprocessing in active sites where each positive and negative charges occur, while high dipole moments would promote efficient processing.

These preliminary analyses suggest several important points on gamma polymerase fidelity: a) the causes for effects of similarity between nucleotides on misrecognition are unknown, structural similarity having effects opposite to those of dipole moment similarities; b) nucleotide properties affecting misrecognition are unknown but correlate with dipole moments; c) separating, even only conceptually, polymerase misrecognition from misincorporation, could be useful to understand polymerase accuracy; d) many misincorporations might be due to spontaneous mutations (with rates proportional to the dipole moment model) in the nucleotide occurring after accurate recognition by the polymerase, but before incorporation, resulting in misincorporation despite accurate recognition; e) alternatively, the polymerase's binding site might have in-built bias for hydrophobic misprocessing.

3. Selection on the gamma polymerase's misincorporation rates

Grantham [20] developed a matrix of dissimilarities based on major physico-chemical properties of amino acids (amino acid composition, polarity and molecular volume) that correlates best with amino acid replacement frequencies. From that matrix, Gojobori et al [17] estimated an average change in amino acid physico-chemical properties due to residue replacements for nucleotide substitutions in protein coding regions (see last line of table 4 in [17]). For example, A<->G substitutions have the lowest average impact, while G<->T have the greatest impact. One expects a negative association between impact on protein structure and the frequency of a nucleotide substitution. For pseudogenes, which do not code for proteins, the correlation between this average impact and the frequency of corresponding mutations (data from [17]) is weak ($r = -0.33$, one tailed $P = 0.15$), and even weaker after differences between transition and transversions have been accounted for ($r = -0.18$, one tailed $P = 0.29$). However, for mutation frequencies in coding sequences, natural selection against dysfunctional proteins has specifically decreased frequencies of non-conservative substitutions, and a strong negative correlation exists between impacts on protein structure and the frequency of a nucleotide substitution ($r = -0.828$, one tailed $P = 0.00044$). Accounting for differences between transitions and transversions does not alter qualitatively this result ($r = -0.749$, one tailed $P = 0.0025$).

Hence different misincorporations by the gamma polymerase [18] affect differently the coding properties of genes. The polymerase probably mainly adapted to avoid high impact nucleotide misincorporations. This can be tested by examining the correlation between misincorporation kds and the amino acid impact distances presented in [17], which will indicate to what extent these misincorporation rates resemble what is expected for pseudogenes (suggesting no selection occurs), or coding genes (suggesting the gamma polymerase is selected to minimize substitution impact on proteins). This correlation is negative, stronger than for pseudogenes, but weaker than for functional genes after selection (r = -0.434, one tailed P = 0.079). Controlling for differences between transitions and transversions does not alter much this result (r = -0.323, P = 0.15). The same holds after accounting for effects of dipole moments (Figure 2) on misincorporation rates: kds decrease with distances between replaced and replacing residues, but results are intermediate between mutation patterns observed for pseudogenes and genes that actually code for proteins (r = -0.44, one tailed P = 0.076).

This indicates that misincorporation rates include an adaptive component that minimizes the potential impact of nucleotide misincorporations on proteins. It is probable that a balance exists between minimizing different misincorporation rates, because the same active site in the polymerase is responsible for them. Hence the misincorporation pattern cannot be adapted to minimize all misincorporation rates, only to optimize misincorporation effects at protein levels. For frequencies of mutations observed in genes, selection affects each site (more or less) independently, hence impacts are minimized, resulting in much stronger correlations between mutation frequencies and impact at the protein level than observed for misincorporation rates, because the same active site produces the various types of misincorporations. The results indicate that this balancing effect due to interactions between different misincorporation types by the same active sites must be relatively strong in the gamma polymerase, otherwise the correlation with amino acid dissimilarities would resemble much more that found for coding genes. The matter of adaptively-tuned misincorporation rates by polymerases is nevertheless an interesting line of research that would gain from being developed further, including along the methods used here.

4. Gene-specific substitution matrices for human mitochondrial protein-coding genes

Misincorporation by gamma polymerases during replication is a major factor causing mutations. This factor is itself influenced by dipole moments of nucleotides, similarities between them, and greater selection pressures against specific misincorporation rates than on other rates (see previous sections). Here I examine observed mutation patterns in human mitochondrial genes.

Numbers of nucleotide substitutions for each of the 12 possible substitutions were counted from tabulations at http://www.mtdb.igp.uu.se/ [1] and http://www.mitomap.org/MITOMAP [2], separately for each gene (Table 1). Values are percentages of sites where a given mutation was observed among all sites where the substituted nucleotide mutated in that gene. The variation in that percentage within a given gene is mainly due to differences between transitions and transversions, the former dominating. Hence for further analyses, for each gene, mean percentages for transitions and transversions were calculated separately and subtracted from the observed percentages for transitions and transversions, respectively. This adjustment excludes effects due to differences between transitions and transversions in mutation percentages observed for each given gene. The two last columns in Table 1 are Pearson correlation coefficients of percentages adjusted for differences between transitions and transversions and adjusted (along the same criterion) kd's of nucleotide misincorporations by the gamma polymerase (s), and after adjusting also for Grantham physico-chemical distances (s'). Correlation coefficients s are positive in 12 among 13 genes, a significant majority of cases according to a one tailed sign test (P = 0.000854). The correlation is significant (P < 0.05) at the level of a single gene for three genes, ND1, CO1 and AT8 (marked by asterisks in Table 1).

Results are only slightly altered after accounting for differences between transitions and transversions. Further analyses (s' in Table 1) using the residual misincorporation rates and the residual mutation percentages, calculated from their regressions with Grantham's amino acid dissimilarities do not change results much. These results show that variation in percentages of mutations of different types is to some extent due to misincorporation by the gamma polymerase, but a large part of the variation between substitution percentages remains unaccounted for. It is probable that natural selection against various mutations occurs, so that percentages in Table 1 are composites of misincorporation rates and other factors, such as selection against specific mutations. However, taking selection into account by using residuals from the regression of mutation frequencies with amino acid dissimilarities does not change patterns much. Hence further major factors affect observed mutation patterns, besides misincorporation rates and selection on coding impacts of mutations (and misincorporations).

5. Effects of deaminations and selection on mutation matrices

If one assumes that large parts of the variation that is not explained by the gamma polymerase's misincorporation rates in the previous analyses is due to selection, one can estimate which types of mutations are more or less prone to selection by analysing the residuals of the adjusted percentages (for each gene) from the regression with misincorporation. The line 'Res' in Table 1 indicates the number of genes for which this residual was positive, meaning that the percentage of that mutation was greater than expected from the regression with misincorporation. For two types of mutations, C->A and T->C, there were 10 such genes, which according to two tailed sign tests yields a significant tendency for observing percentages greater than expected by misincorporation (P = 0.046) as indicated by P in Table 1. Hence C->A and T->C are more frequent than expected by misincorporation. At least for T->C, there are two plausible explanations. T->C is a transition, and transitions cause relatively little functional effects at the level of coding properties of codons, suggesting low counterselection, hence relative over-representation (positive residuals). This explanation does not seem adequate, because the effect is not strong for other transitions (A->G, G->A and even opposite for C->T, where residuals were positive for only 2 genes (P = 0.0095, two tailed sign test). The latter effect on C->T is however also compatible with the second explanation for T->C. Deamination, promoted by single strandedness during replication, contributes to A->G mutations on the mitochondrial heavy strand DNA, which corresponds to T->C in Table 1 which uses the complementary light strand DNA annotation. Hence the systematic excess in T->C and systematic lack of C->T would be due to a factor that does not relate to misincorporation by the gamma polymerase, nor to selection, but presumably to the replicational mutation gradient of A->G. Residual analysis also indicates systematic underrepresentation of a further mutation type, G->T (P = 0.0095, two tailed sign test), a transversion that might be particularly counterselected [17]. Indeed, numbers of positive residuals tend to decrease with mean physico-chemical distances between replaced and replacing amino acids associated with these nucleotide mutations (r = -0.38, not statistically significant).

Gene	A		C		G		T		A-C	A-G	A-T	C-A	C-G	C-T	G-A	G-C	G-T	T-A	T-C	T-G	s	s'
ND1	272	116	344	124	112	45	228	48	2.8	88.9	8.3	13.3	4.7	82.0	87.8	6.1	6.1	7.7	78.2	14.1	55*	53*
ND2	326	114	349	109	99	33	268	77	3.8	91.5	4.7	15.3	5.9	78.8	80.5	9.8	9.8	3.3	89.1	7.6	-5	-11
CO1	419	121	462	121	250	59	410	97	5.7	89.5	4.8	12.7	4.2	83.1	91.8	3.3	4.9	6.7	90.4	2.9	55*	47
CO2	196	65	214	59	102	39	172	55	7.7	83.1	9.2	4.3	10.0	84.7	90.2	7.3	2.4	9.4	85.9	4.7	13	6
AT8	80	42	69	31	13	9	45	26	0.0	95.2	4.8	11.1	0.0	88.9	100	0	0	0	92.9	7.1	74*	71*
AT6	206	115	230	81	71	47	174	95	4.2	90.8	5.0	8.1	5.8	86.2	90.2	7.8	2.0	3.0	87.0	10.0	40	43
CO3	210	87	249	70	116	44	209	69	11.6	84.9	3.5	6.3	3.8	90.0	93.2	4.6	2.3	9.5	87.8	2.7	45	37
ND3	102	41	102	27	37	13	105	29	18.4	73.7	7.9	13.8	6.9	79.3	84.6	15.4	0	5.1	92.3	2.6	17	3
ND4l	84	27	92	19	36	12	85	23	7.7	84.6	7.7	7.7	0	92.3	84.6	0	15.4	8.7	78.3	13.0	20	47*
ND4	416	144	473	133	137	32	352	92	10.8	84.2	5.0	10.2	5.7	84.1	87.9	12.1	0	5.6	87.9	6.5	31	15
ND5	518	207	580	183	190	49	416	117	7.8	85.5	5.7	18.7	18.7	62.6	85.6	8.1	6.3	4.9	90.9	4.2	19	3
ND6	198	53	187	72	37	21	103	32	4.8	82.5	12.7	13.4	10.5	76.1	90.9	9.1	0	3.6	92.7	3.6	27	-25
Cytb	326	142	391	141	137	67	287	95	4.3	88.7	7.0	8.6	6.2	85.2	89.7	10.3	0	1.9	92.5	5.7	12	-1
Res									4	7	7	10	5	2	6	7	2	5	10	5		
Dssh									-29	16	3	38	23	-39	-13	30	-17	-63*	35	-9		
Pos 1									-12	17	-11	15	4	-20	-6	12	-7	23	-18	2		
Pos 2									-3	-1	11	10	-15	4	-1	1	-1	-27	13	4		
Pos 3									-11	-23	53*	29	33	-35	-56*	52*	20	-39	28	-1		
Dloop									22	-39	27	-9	42	-26	20	30	-41	-31	53*	-42		
Pos 1									-13	19	-11	5	2	-7	49*	-18	-54	30	-3	-19		
Pos 2									9	-20	36	41	-37	-4	-19	13	13	-21	33	-34		
Pos 3									39	-50	38	-9	43	-20	-40	60*	-26	-46	53*	-26		

Table 1. Percentage of mutations observed in each human mitochondrial protein coding gene. A, C, G, T indicate the number of that nucleotide in that gene, followed by the number of sites with that nucleotide that are polymorphic. 's' is the Pearson correlation coefficient of percentages adjusted for differences between transitions and transversions and adjusted nucleotide misincorporations by the gamma polymerase (* indicates P < 0.05). The last lines (from Res on) and s' are explained in the text.

6. Mutation gradients across mitochondrial genomes

The previous section indicates that some mutations might be systematically more frequent than expected by misincorporations by the gamma polymerase, and suggests that mutations due to replicational deamination gradients could cause this effect. The study of mutational gradients has used different methods to compare mutation rates at different locations in the genome. Some studies infer mutation rates from phylogenetic comparisons among species of nucleotide contents at given sites (i.e. [13, 14]). Phylogeny-inferred kinetics for A->G and C->T gradients match the properties of the underlying chemical processes: the chemically faster C->T deamination saturates faster in computational analyses with duration spent single stranded than the slower A->G reaction [13, 14, 21]. Other studies infer mutation rates from gene nucleotide contents: for the C->T deamination, one expects relatively high C and low T contents in regions close to replication origin(s), and the opposite for genes with high durations spent single stranded [11, 12, 15, 21, 22, 23].

The method used here is closer to direct observation of mutations, because it compares only between genomes from the same species (*Homo sapiens* in this case). This means that one is closer to an 'instantaneous' observation of mutations. This procedure decreases numbers of undetected multiple changes. I did not use a full phylogenetic model of all human mitochondria to infer mutation rates. Data in Table 1 are for a simplified procedure that counts numbers of sites within a gene where a given type of mutation was observed and calculates the percentage of sites with that nucleotide where that mutation occurred, assuming that the most common nucleotide at any given site is the ancestral nucleotide.

Durations spent single stranded are calculated as previously [11, 12, 21, 22, 24]. I explored for replicational and transcriptional gradients (Dssh and Dloop in Table 1) for each of the 12 mutations, not only for A->G and C->T. This is because time spent single stranded might also affect other mutations, notably transversions [13]. Correlational analyses for gradients (analysis across rows, one per column in Table 1) used the residuals of mutation rates from their regression with misincorporation rates (residual analysis is across columns, one regression calculated per gene/row in Table 1), in order to exclude effects of polymerase inaccuracy on mutational gradients. However note that using the raw mutation percentage data as in Table 1, gradient analyses do not change much.

Two potential gradients in duration of singlestrandedness are considered, singlestrandedness during replication and during transcription (indicated in Table 1 by Dssh and Dloop, respectively). The last rows in Table 1 show Pearson correlation coefficients between residual mutation percentages and times spent single stranded during replication (Dssh) and transcription (Dloop). The hypothesis of singlestrandedness expects positive correlations, but this was observed only for half the cases, for each replication and transcription. There was a significant drop in T->A mutations along the replicational gradient, and a significant increase in T->C mutations along the transcription gradient (Figure 3). The latter effect is predicted by deamination gradients. Deamination gradients are also expected for G->A, but were not observed. Data in Table 1 only support the hypothesis of a deamination gradient for T->C. They cannot differentiate between replicational and transcriptional gradients. It is notable that in this case the predicted G->A gradient is not stronger than gradients observed for other mutations. Apparently, another mutation, T->A, reacts to single strandedness, but in the direction opposite to that expected (singlestrandedness is predicted to increase mutations, not decrease them). Other, less direct methods based on phylogenetic reconstructions, perhaps fail to detect this gradient because selection, at larger evolutionary scale, might have weeded out many mutations such as the transversion T->A (this type of mutation implies non-conservative changes at the amino acid coding level), leaving mainly neutral and close to neutral ones. Indeed, transitions affect less coding properties than transversions (transitions cause on average more conservative amino acid changes than transversions). This would explain why phylogenetic comparisons detected weaker signals for transversion than transition gradients, while analyses in Table 1 for almost instantaneous mutations are apparently less affected by natural selection occurring after a mutation happened and do not show differences in gradients between transitions and transversions. These comparative data restricted to *Homo sapiens* confirm only the (heavy strand) deamination of A->G (corresponding to T->C in the annotation used here) at the level of a transcriptional gradient.

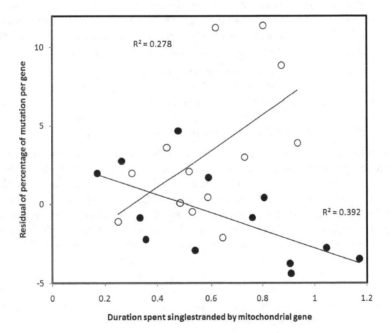

Figure 3. Mutations versus singlestranded during replication (T->A, filled symbols) and transcription (T->C, circles). Mutation percentages are residuals from regressions with misincorporation by the gamma polymerase, calculated based on data from Table 1.

The results suggest that mutation rates estimated from sequence comparisons within a single species reflect misincorporation rates, but barely confirm well established observations of deamination gradients, which were based on comparisons between evolutionary more distant sequences, and on nucleotide contents of single sequences. Apparently, instantaneous mutation rates reflect misincorporation by gamma polymerase, while the effects of deamination gradients, which result from a biased cumulation of mutations, might result from long term processes and are therefore more detectable at a wider evolutionary scale.

7. Mutation gradients and selection at different codon positions

The issue of effects of selection on mutational gradients can also be investigated by analysing separately codon positions, as indicated for replicational and transcriptional gradients in Table 1. In terms of replicational gradients, there were no gradients detectable for any mutation at first and second codon positions, but there were three gradients, one negative (G->A) and two positive (A->T and G->C) at third codon positions. Hence these analyses confirm that replication gradients are more detectable where the mutation is synonymous or has little impact because causing a conservative amino acid change, as occurs at third codon positions, but not or much less at first and second codon positions. However, the specifically predicted deamination gradients are not detected. The opposite is observed for G->A (corresponding to C->T mutations on the heavy DNA strand), this mutation unexpectedly decreases along the singlestrandedness gradient, while an increase was expected.

Assuming a transcriptional gradient in singlestrandedness, the expected positive G->A gradient is detected for first codon positions. This is the only statistically significant gradient detected that is not at third codon position. The transcriptional gradient analyses at third codon position confirm the gradient observed for pooled codon positions for T->C, which fits the deamination gradient, and detects a gradient for G->C mutations.

Comparing the absolute values of the correlation coefficients in Table 1 for replicational and transcriptional gradients, correlations are stronger with transcriptional singlestrandedness, however this analysis does not account for the expected positive direction of the correlations of mutations with singlestrandedness. If one assumes that correlations should be positive (singlestrandedness should increase mutations), one does not detect any systematic difference between replication and transcription. The human mutation data might be better explained by transcriptional singlestrandedness, but the matter remains unclear. Deamination gradients are more detectable assuming transcriptional than replicational singlestrandedness, suggesting that deaminations observed in human sequences occurred mainly during transcription. The fact that more gradients are detected at third codon positions than at other positions indicates that selection against mutations affecting protein structure occurs and prevents detecting mutational gradients due to singlestrandedness.

8. Mutation gradients and misincorporations

Analyses in the previous section suggest that mutational gradients exist in mitochondria, but are less detectable at the evolutionary scale reflected by sequence variation within *Homo sapiens* populations than when comparing between evolutionary more divergent sequences belonging to different species. Nevertheless, additional analyses show that replicational gradients confound effects of misincorporation by the gamma polymerase. Indeed, the column 's' in Table 1 shows that while mutation patterns in most genes overall fit the pattern predicted by misincorporation, this extent varies widely among genes (from -5 for ND2 to 73 for AT8). My first guess was that gene size (from 69 to over 600 codons, for AT8 and ND5, respectively) differences cause this. My assumption was that estimations of mutation patterns are less accurate in short genes, causing low correlations (low s) between observed mutation patterns and misincorpration rates. However, if this was true, one would expect a better match with misincorporation patterns in long genes, but surprisingly, patterns fit best in AT8: sampling inaccuracy does not explain variation in 's'.

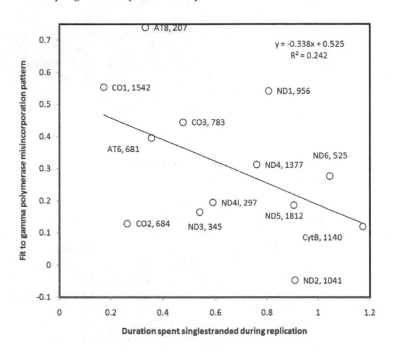

Figure 4. s from Table 1 as a function of singlestrandedness during replication. Mutation patterns resemble those predicted by misincorporation by the gamma polymerase in genes that remain singlestranded for a short time during replication. Values indicate gene lengths.

Replicational mutation gradients might explain variation in s between genes: mutation patterns in genes that endure short periods of singlestrandedness during replication should be least affected by replication gradients, and fit best the pattern predicted by gamma polymerase misincorporation, and vice versa (Figure 4). Indeed, s decreases with singlestrandedness during replication (r = -0.49, P = 0.045, one sided test; but there was no correlation of s with singlestrandedness during transcription, r = -0.27, P < 0.10).

Inaccurate 's' estimation due to short genes affects results in Figure 4. Short genes fit less the trend in Figure 4 than large genes (gene size is indicated in Figure 4): absolute values of residuals calculated from the regression in Figure 4 decrease with gene size (r = -0.45). Hence 21 percent of variation in s unexplained by singlestrandedness is from sampling effects. Accounting for them, the correlation in Figure 4 is r = -0.63. This means that sampling effects affect less 's' (and estimates of observed mutation rates) from Table 1 than singlestrandedness. This stresses the importance of mutational gradients despite weak results in Table 1.

Singlestrandedness during replication is an even better predictor of the fit between observed mutation patterns and gamma polymerase misincorporation when residual analyses account for each Grantham distances between replaced and replacing amino acids (s' in Table 1). This s' decreases more than s with replicational singlestrandedness (r = -0.6277, one tailed P = 0.011). Interestingly, using transcriptional singlestrandedness yields r = -0.468 (one tailed P = 0.053). Accounting for total singlestrandedness during both replication and transcription by summing both up and analysing the correlation of s' with this sum of replicational and transcriptional singlestrandedness yields r = -0.649 (one tailed P = 0.0083). In each of these analyses using s', gene size had a significant impact on residuals. Accounting for that effect systematically increased correlations between s' and replicational, transcriptional, and the combination of both singlestrandedness (r = -0.811, r = -0.68 and r = -0.89).

9. Mutation patterns: effects of dipole moments or gamma polymerase misincorporations?

Figure 2 shows that even after accounting for differences between transitions and transversions on misincorporation rates, differences between dipole moments of the substituted and the substituting nucleotides explain part of the variation in misincorporation rates. Hence both factors (dipole moment or misincorporation by the gamma polymerase) are confounded, and one cannot be sure which affects mutation patterns, or whether they affect each independently observed mutation patterns. For that reason I calculated residuals of adjusted kds from the regression with signed differences in dipole moments (data from Figure 2) and calculated correlations between these residuals and the adjusted mutation percentages (calculated from Table 1) for each gene. This version of s is adjusted for effects of dipole moments on misincorporation rates, and is positive for all genes. This adjusted s increased as compared to s from Table 1 in 8 (and decreased in 5) genes. Hence adjusting misincorporation for effects of dipole moments only slightly increases its fit with observed mutation patterns.

However, when examining the increase in s after adjusting for dipole moment effects in relation to replicational singlestrandedness, this increase is proportional to singlestrandedness during replication (not shown). This suggests that effects of the component of misincorporation that is independent of dipole moments increase with singlestrandedness. Hence singlestrandedness interacts with gamma polymerase fidelity. In these analyses, this fidelity is separated into a component associated with dipole moments, and a different component.

According to the analyses, it is the latter, unknown factor that increases its effects on ob-served mutation patterns with singlestrandedness during replication.

Similar residual analyses for dipole moments show that observed mutation patterns do not fit well with differences in dipole moments after calculating residuals from their regression with misincorporation rates (these analyses inverse between dependent and independent in Figure 2). These correlations were negative in 11 among 13 genes, suggesting a weak effect that is op-posite to that expected by the hypothesis that mutations decrease dipole moments [4].

The latter analyses indicate that dipole moments affect mutation rates through their effects on misincorporation by polymerases, but not directly on spontaneous alterations of single strand-ed DNA. Misincorporation by gamma polymerase has at least two components, one related to dipole moments, and another one, unrelated to dipole moments. Effects of the latter on muta-tion patterns increase with singlestrandedness. Analyses in a previous section suggested that distinguishing between misincorporation due to nucleotide misrecognition versus misincor-poration due to nucleotide alteration after accurate recognition could prove valuable. It is not clear whether the effect independent of dipole moments that increases with singlestranded-ness relates to misrecognition, alteration after recognition, or a subcomponent of any of these. Hydrophobic bias (for low dipole moment) in relation to misincorporations by the gamma pol-ymerase binding site for nucleotides is not explained by a simplistic analysis of the residues composing the active site of gamma polymerase. Nevertheless, these results indicate that the 'age' of the replication fork has some effect on its fidelity.

A similar comparison can be done between s and s' in Table 1. Here one sees that s', as com-pared to s, is lower than s in 11 among 13 genes. Hence gene-specific mutation patterns match misincorporation rates after accounting for differences between transitions and trans-versions better than after accounting, in addition, for selection against non-conservative amino acid replacements resulting from nucleotide substitutions: s' as compared to s de-creases with singlestrandedness. Hence in this case, accounting for selection against non-conservative amino acid replacements improves slightly the match of observed mutations with misincorporations for genes with short singlestranded exposure, but mainly decreases that match for those with long singlestrandedness. This effect is opposite to the one reported in a previous paragraph for accounting for dipole moment effects. Accounting for the latter improves the match between mutation and misincorporation patterns with singlestranded-ness, while accounting for selection decreases that match.

10. More evidence for complex indirect effects of mutation gradients

Sampling inaccuracy might affect estimates of s and s' in Table 1. A further potential indirect factor with opposite effect might exist. Duration spent single stranded used is for a gene's mid-point, a good approximation for short genes, but increasingly inaccurate the longer the gene. In order to evaluate this, absolute residuals of mutation percentages (Table 1) from their regres-sions with misincorporation rates (both adjusted for differences between transitions and trans-versions) are plotted versus numbers of potential sites that could mutate for that mutation type in that gene (Figure 5). Residuals tend to decrease with sample sizes up to approximately 250 nucleotides, which corresponds to an average of a sequence of 1000 base pairs (for each mutation type, there is only one substituted nucleotide, so on average, these mutations occurred over a total sequence that is about four times longer). The absolute value of residual mutation percentages increases with sample size up from about 250 nucleotides.

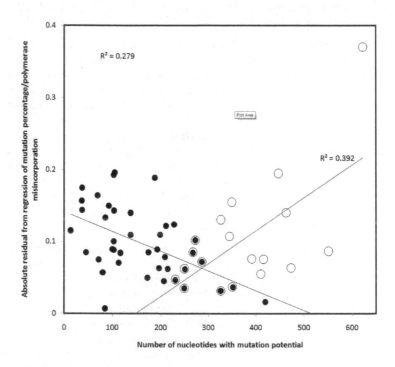

Figure 5. Residual mutation percentage (absolute value) from regression with gamma polymerase kds for each nucleotide in each gene, versus numbers of potentially mutating nucleotides. The decrease indicates a sampling effect: samples up to 200-300 nucleotides fit better misincorporation patterns because of sampling effects. Beyond 250, inaccuracy increases, perhaps because different gene regions have different mutation regimes.

The decreasing pattern is what one expects from sampling effects: up to about 1000 base pairs, longer genes enable to estimate better mutation patterns (the absolute residual is small). But for genes longer than that threshold, absolute residuals increase, hence mutation patterns tend to fit less well misincorporation as predicted by the gamma polymerase. This could be due to the mixing of regions with different singlestrandedness, which perhaps alters non-linearly mutation patterns. A similar effect where estimation inaccuracy of mutation rates decreases, then increases with sequence length exists for the correlation between rates of morphological and molecular evolution [25]. The threshold was for sequence lengths around 1200 base pairs, indicating that estimates of mutation rates (mainly from vertebrate mitochondrial protein coding sequences, as those analysed for *Homo sapiens* here) decreased beyond that sequence length. It was suggested, as for Figure 5 here, that mutation patterns change with the relative position of a gene, and that for long regions, more than one mutation regime might be mixed, decreasing the accuracy of analyses. Figure 5 follows that principle, and indicates a similar threshold.

11. Mutational gradients after accounting for amino acid replacement impacts on proteins

Previous sections show that indirect effects of gradients in singlestrandedness on mutation patterns exist (i.e. Figures 4 and 5). Yet analyses of mutation percentages, or mutation percentages after accounting for differences between transitions and transversions, and after accounting for effects of misincorporation by gamma polymerase, do only marginally enable to detect mutation gradients with singlestrandedness, and this for any codon position. The analyses of gradients that separate codon positions indicate that natural selection might affect mutation patterns (Table 1), and could mask mutational gradients according to singlestrandedness. Selection against non-conservative amino acid replacements also affects mutation percentages. Analyses for singlestrandedness gradients did not yet account for that latter factor, in addition to misincorporation by the gamma polymerase and differences between transitions and transversions.

I calculated residuals of mutation percentages (adjusted for differences between transitions and transversions) from their regression with mean physico-chemical (Grantham's) distances between replaced and replacing amino acids resulting from that nucleotide substitution in coding sequences (for mutation percentages across all codon positions), separately for each of the protein coding genes. Hence this analysis is across columns, for each row in Table 1. Then, for each substitution type, I calculated correlations with singlestrandedness during replication, transcription, and their sum (these analyses are across rows, for each column, on residuals produced by the latter 'row' analysis across columns). This yields correlations between residual mutation rates and singlestrandedness for each mutation type. The majority of these are positive correlations (Table 2): mutation percentages (after accounting by residual analyses for differences between transitions and trasversions, misincorporation rates and Grantham distances (assumed to reflect selection against dysfunctional proteins)) increase with singlestrandedness during replication (11 among 12 cases, exception A->C mutations), transcription (11 among 12 cases, exception A->G mutations) and their sum (all cases). Hence overall, singlestrandedness promotes all types of nucleotide substitutions, not only deaminations A->G and C->T, for both replicational and transcriptional singlestrandedness. Their sum improves correlations in half the cases. Correlations were statistically significant (one tailed $P < 0.05$) for one correlation with replicational singlestrandedness (T->C), two with transcriptional singlestrandedness (C->G and T->C) and three with the sum of both (A->T, T->C and T->G).

Correlations were stronger with transcriptional singlestrandedness than replicational singlestrandedness in 7 among 12 cases, which does not indicate which among the two is the most important factor. Possibly, singlestrandedness during replication and during transcription affect differently different substitution types, or differences are random. These analyses clearly show that after accounting for mean effects of substitutions on proteins, percentages of all types of nucleotide substitutions increase with singlestrandedness during each replication and transcription. These clear patterns were not detectable without accounting for mean nucleotide substitution impact on physico-chemical properties of coded amino acids. It seems these effects prevented detecting mutation gradients for substitutions that were not deaminations. Singlestrandedness increases at least slightly probabilities of all types of substitutions.

Substitution	Rep	Trans	Both
A->C	-0.159	0.401	0.066
A->G	0.428	-0.242	0.241
A->T	0.400	0.447	0.481*
C->A	0.378	0.063	0.291
C->G	0.319	0.468*	0.433
C->T	0.243	0.315	0.312
G->A	0.091	0.339	0.216
G->C	0.395	0.322	0.421
G->T	0.425	0.340	0.448
T->A	0.301	0.381	0.382
T->C	0.478*	0.531*	0.573*
T->G	0.452	0.350	0.469*

Table 2. Pearson correlation coefficients of time spent singlestranded during replication, transcription, and their sum versus substitution percentages in the 13 human mitochondrial protein coding genes adjusted for differences between transitions and transversions, misincorporation rates and for mean effect of the substitution on Grantham's physico-chemical distances between replaced and replacing amino acids.

Causes for differences in gradient strengths for different substitution types are not known. Gradients are strongest for substitutions involving a small absolute change in nucleotide dipole moment, and weakest for those where the absolute change in dipole moment is large. Speculatively, large dipole differences may affect even when singlestrandedness is short, so that no strong gradient is detectable, because the main effect is the dipole moment, independently of singlestrandedness. For small dipole moment differences, the dipole moment effect woult hence be enhanced by singlestrandedness, resulting in a gradient.

12. Nearest neighbour effects on mutation rates

Previous analyses of mutation patterns in human mitochondrial protein coding genes fit expectations according to several factors: misincorporation by the gamma polymerase, selection against mutations that alter amino acid properties, and dipole moments of nucleotides. A hierarchy between these factors exists. In addition, they interact: misincorporation rates are also affected by selection against non-conservative mutations; and gradients in single-strandedness affect extents by which the various factors affect mutation patterns. Only after adequate accounting for misincorporation and selection (and differences between transitions and transversions), mutation gradients along durations of singlestrandedness are cleary observed for all types of nulceotide substitutions.

Flank		5'						3'					
A		Tot	Mut	A	C	G	T	Tot	Mut	A	C	G	T
	A->	968	319		15	290	14	928	238		19	193	26
	C->	1037	289	22		10	257	1069	363	21		12	330
	G->	461	166	150	11		5	371	89	79	6		4
	T->	897	278	8	244	26		1275	470	18	395	57	
C													
	A->	1097	326		36	260	30	1063	447		22	401	24
	C->	1285	327	41			270	1293	390	35		34	321
	G->	311	114	100	8		6	505	203	175	20		8
	T->	1110	245	23	203	19		981	286	17	240	29	
G													
	A->	378	158		6	145	7	447	229		12	204	13
	C->	503	156	15		28	113	322	154	21		8	125
	G->	254	72	60	8		4	256	72	68	2		2
	T->	227	83	6	67	10		323	115	6	102	7	
T													
	A->	890	425		18	373	34	888	303		20	271	12
	C->	823	252	19		21	212	1102	252	57		14	181
	G->	328	137	123	8		3	222	124	111	10		3
	T->	676	295	13	264	18		668	198	13	171	14	
	A				40	-54	1				72*	40	48*
	C					-37	36					43	60*
	G					51*							91*

Table 3. Dinucleotide sites and mutating sites in human mitochondrial protein coding sequences, separating 5' and 3' nucleotide identity. Last 3 lines are correlations, see text.

Despite the relative complexity of factors described and affecting mutation patterns, this is not an exhaustive list of effects on mutation rates. Notably, nearest neighbour effects exist [26], where identities of nucleotide(s) flanking the mutating site affect mutation rates, as indicated by the editor of this volume after reviewing a former version of this chapter. G and C, the nucleotides with the highest dipole moments, seem to increase mutation rates in various organisms along similar patterns [26-30]. This suggests a physico-chemical basis for nearest neighbour effects, possibly along the lines of dipole moment effects and the stability of DNA duplexes surrounding the mutating nucleotide [26]. These biases are strong enough to justify the need of incoporating at least the strongest nearest neighbour effect in models designed to detect natural selection on mutations [31], which is not surprising as CpG dinucleotides are disproportionately represented among sites with pathogenic polymorphisms [32,33]. Moreover, nearest neighbour effects interact with gene location and the frequency of transcription, suggesting interactions with singlestrandedness [34, 35]. Nearest neighbour analysis of mutation patterns requires large sample sizes, and therefore is unfortunately incompatible with a gene by gene analysis as a function of singlestrandedness in the context of this mitochondrial dataset.

However, even after pooling mutation data from all genes, one would ideally examine the twelve substitutions in relation to each of the 16 combinations of nucleotides at the 5' and 3' positions. Such detailed analyses are also not possible with this dataset. Nevertheless, as known to this author, nearest neighbor effects have not yet been examined in the context of mitochondrial genomes, hence even simplified analyses pooling mutations from all genes and codon positions together may still be valuable. In addition, most nearest neighbour analyses examined do not analyse substitutions in relation to their direction (they pool X->Y with X<-Y), but this can be done on this dataset. Mutation data from all genes and codon positions were pooled, and analysed each time separately in relation to the identity of their 5', and their 3' flanking nucleotide. This yields reasonable samples, and the mutation patterns can be compared according to the different flanking nucleotide identities (Table 3).

The data in Table 3 enable a number of different analyses, only one is presented here, though many others are of interest. For example, biases exist in terms of dinucleotide frequencies, between 5' or 3' flanking by the same nucleotide. I focus here on the analysis of mutation patterns. Numbers in each row in Table 3 were divided by the number of mutating sites among all possible dinucleotide sites for that category (Mut). The column Tot in Table 3, which indicates the total number of dinucleotide sites found independently of the occurrence of a mutation at that site, is indicated but not used in further analyses. The substitution matrices that result are very similar, comparing 5' and 3', and different nucleotide contexts. This is because the overwhelming majority of the variation in mutation rates is due to the difference between transitions and transversions. For that reason, effects of transitions versus transversions were accounted for by subtracting observed mutations rates from the average for transitions and transversions, respectvely, as done in previous analyses. Then these data adjusted for differences between transitions and transversions are compared between different substitution matrices, so that effects of the difference between transitions and transversions is accounted for before comparing the matrices with different neighbours.

The three last lines in Table 3 show Pearson correlation coefficients (x100) between these mutation patterns (adjusted for differences between transitions and transversions). Even after accounting for differences between transversions and transitions, substitution patterns across different 3′ neighbouring nucleotides resemble each other: all six correlations are positive, 4 among these are statistically significant (P < 0.05, one tailed tests because positive associations are expected, see asterisks in Table 3). 3′ G and T had most similar patterns. Hence grouping of mutation patterns according to 3′ nucleotides does not follow purine/pyrimidine nor dipole moment differences. 3′ G seems to affect most mutation patterns, hence results are probably not random also for 3′ nearest neighbour effects.

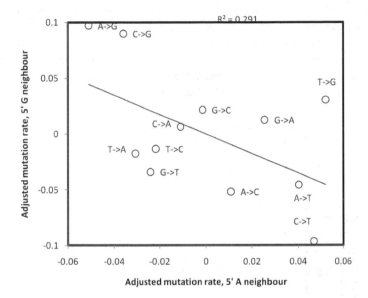

Figure 6. Mutation rates adjusted for differences between transition and transversions for 5′ A and G neighbours in human mitochondrial protein coding genes.

The same analysis for 5′ flanking nucleotides reveals a similar, more enhanced situation. Here, the only statistically significant association is between mutation patterns with 5′ G and T as nearest 5′ neighbour. The weak positive correlations in the 3′ context are negative in the 5′ context, one being close to statistically significant (the comparison between 5′ A and G, see Figure 6): 5′ G affects mutation rates in a way that tends to be systematically opposite to what is observed in other contexts, so that relatively high mutation rates become relative low, and vice versa. Effects of 5′ G on mutation rates are expected, considering previous reports. However, these have mainly shown effects on C->T mutations. The results here show that 5′ G has a systematic effect on all mutation types, some increasing, as expected, but others decreasing in the 5′ G context.

It is notable that the correlation matrices for 5′ and 3′ contexts (in the 3 last lines of Table 3) are very similar, if not in their values, but in their pattern: the ranks, from least to most positive correlation coefficients, are identical (Figure 7). This means that the same effects are at work for 5' and 3' flanks, but that effects are stronger for 5' flanking nucleotides. In this context, it is important to remember that the annotation used here is that of the light strand DNA in the mitochondrion, which bears the coding sequence of most genes. In the elongating light DNA strand, the 3' nucleotide is already present before the mutating nucleotide is added, while the 5' nucleotide is not yet there, and could not possibly have any effect. This is not compatible with a 5' effect during replication, unless one considers that the effect is from the neighbouring nucleotide on the template heavy strand DNA. In that case, the inverse complement would have the major flanking effects, with the strongest effect by the nucleotide that is not yet complemented by the nascent strand (the 5' of the light strand becomes the 3' in the heavy strand), and a weaker but similar effect by the neighbouring nucleotide that is already complemented by the replication process. Along that scenario, neighbouring nucleotides would affect misincorporation rates. This scenario would be very compatible with electrostatic effects, due to dipole moments.

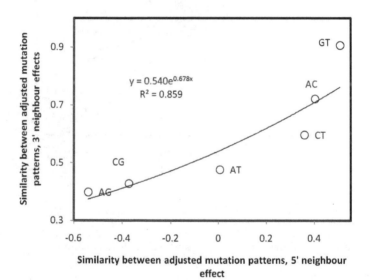

Figure 7. Similarities (Pearson correlation coefficients in the three last lines of Table 3) between transition versus transversion adjusted mutation patterns for 3' neighbouring nucleotides as a function of similarities for mutation patterns found for 5' neighbouring nucleotides (see Table 3). Letters near datapoints indicate the neighbouring nucleotides whose mutation patterns are compared.

It is notable that the 5' G mutation pattern is very similar to the 3' C mutation pattern as these are observed for the light strand ($r = 0.87$). These are the most similar mutation patterns found when comparing 5' and 3' mutation patterns. Because 3' C on the light strand is 5' G on the heavy strand, this similarity indicates that the factor at work involves both strands, always involving the 5' G nucleotide.

Alternative explanations not involving effects on misincorporation rates, such as dipole moments and 'spontaneous' (non-enzymatic) mutations are also very plausible. The latter are more compatible with the similarities in patterns between 5' and 3' effects and effects on both strands, but less with the strong directional effect detected (less similar mutation patterns between 5' than 3' substitution patterns).

Hence strong neighbouring effects are detected on mutation patterns observed in human mitochondrial genomes, yet their cause remain unknown, and might, as for other effects on mutation patterns, have different physico-chemical causes, combined with some biological factors.

13. Dipole moments and retrotranscription rates by the gamma polymerase

The various analyses described show complex effects, most of them are confirmative of phenomena that have already been described. Indeed, it is quite trivial that misincorporation rates affect mutation frequencies, and that these frequencies are decreased by selection against dysfunctional proteins. Gradients in singlestrandedness as affecting mutation rates are also known, though the fact that they affect all or most mutation types is relatively original to the analyses presented here. A similar rationale relates to the original component of the results from the nearest neighbour analyses. However, the fact that so many different factors are jointly considered in the analysis of a single dataset of mutations is not the sole major originality in terms of potential mechanisms explored in this chapter.

The hypothesis that dipole moments affect misincorporation, mutation rates, and mutation gradients, is a major potential novelty. Unfortunately, when its effects are detected, these are not well understood: the main effect on misincorporation is that of absolute dipole moment change, and the bias favouring low dipole moments remains unexplained.

The suggestion by David Stuart, the editor of this volume, to examine associations between dipole moments and elongation rates in relation to the inserted nucleotide [36] yields interesting results in this respect, confirming that dipole moments affect the incorporation rate of nucleotides into nascent DNA. Results below indicate complex mechanisms, and should be considered as preliminary and with extreme caution. First, it seems that kms of incorporations of nucleotides increase with dipole moments, as one would expect if high dipole moments enable quick processing by the (charged and hydrophilic) active site of the gamma polymerase, though this effect is not statistically significant at $P < 0.05$.

However, electron singlets and or triplets of molecules can be in an 'excited' state, which modifies the dipole moment of the molecule, as calculated by Bergmann and Weiler-Feilchenfeld [6] (therein table VII) for nucleotides. Dipole moments for the excited triplet state correlate positively with nucleotide insertion rates ($r = 0.9865$, $P = 0.007$, one tailed test). Considering that more than one correlation test was done (for the regular and the two excited dipole moments), this result is not statistically very strong (especially that only 4 datapoints are involved in the analysis).

I assume that implicitly, the hypothesis developed by the editor, following my initial interest in effects of dipole moments on polymerase activity, is that if dipole moments affect nucleotide incorporation rates, discrimination against incorporation of the much more common ribonucleotides should associate negatively with (deoxyribo)nucleotide dipole moment. Indeed, the activity of the gamma polymerase as measured by Kasiviswanathan and Copeland [36] includes also the (mis?) incorporation rates of ribonucleotides on the template of DNA, and these associate negatively with nucleotide triplet excited dipole moment (r = -0.963, one tailed P = 0.019). In addition, the rate of reverse transcription by the gamma polymerase, where deoxyribonucleotides are inserted on the template of (mis)inserted ribonucleotides correlates positively with the mean of the singlet and triplet excited dipole moments (r = 0.99984, one tailed P = 0.00008, see Figure 8).These analyses yield notable results, though they are not necessarily as statistically robust as they seem, due to the low number of degrees of freedom (only four datapoints). In addition, correlations between each of the kms and each the dipole moment, the singlet and excited, and their average was calculated, in total 12 correlations. In these cases, according to a strict Bonferroni criterion to correct for multiple testing, to keep P <0.05 while testing 12 times a hypothesis, one should use the threshold of P = 0.05/12= 0.0042. According to that often overconservative criterion, only the result in Figure 8 remains statistically significant.

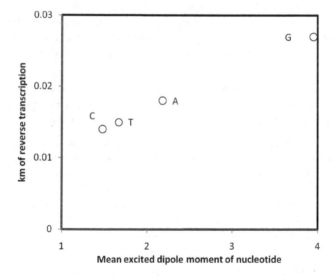

Figure 8. Rate of deoxyribonucleotide 'reverse' incorporation as a function of its mean excited dipole moment on the template of ribonucleotide.

The data nevertheless confirm the hypothesis that nucleotides are processed on the basis of their dipole moment, where nucleotides with high dipole moments are more rapidly correctly processed. This result might actually explain also the results obtained in earlier sections on associations between dipole moment changes and nucleotide misrecognitions. The rate of a process, and its accuracy are frequently negatively associated. Hence if correct incorporation is proportional to dipole moments, misincorporation might be (as observed) inversely proportional, explaining that patterns in Figures 1 and 2 are opposite to predictions: the hydrophilic active site will handle correctly more rapidly a nucleotide with high dipole moment, and more probably mishandle a nucleotide with low dipole moment.

14. General discussion

The analyses presented above show that mutation patterns estimated from the simple comparison between sequences from a species confirm the patterns expected from experimentally determined misincorporation rates for the gamma polymerase. This is an important confirmation that comparative analyses yield trustable estimates of mutation patterns and rates. Analyses support, to lesser extents, that mutation patterns across genes are determined by durations spent singlestranded, and suggest that in order to detect such effects, comparisons involving longer evolutionary time spans than those implied by separations between different individuals from a single species are required to detect the cumulation of mutations due to singlestrandedness.

Grantham's physico-chemical distances between replaced and replacing amino acids affect misincorporation rates by the gamme polymerase, and percentages of mutations observed in protein coding genes. This suggests that natural selection to conserve protein function affects each of these two different patterns. Gradients of mutations with singlestrandedness are barely detectable without controlling for effects of Grantham distances on mutation percentages, several indirect effects are observable that indicate interactions between singlestrandedness and misincorporation patterns by the gamma polymerase. After the effects of Grantham distances on mutation percentages are accounted for by residual analyses, the expected increase in mutation percentages with singlestrandeness becomes detectable in all types of substitutions. This is a notable result, because singlestrandedness was believed until now to affect only or mainly substitutions due to deaminations (A->G and C->T). Analyses do not succeed to indicate which of replicational and transcriptional singlestrandedness is most relevant to predict mutations. Further chemical processes accounting for effects of singlestrandedness on mutation types besides deaminations have to be investigated and suggested.

It seems that gamma polymerase misincorporation patterns change with single strandedness, which may reflect the duration of activity by the replication fork's molecular 'machinery'. Only molecular experiments much more developed than those used until now could yield such results. This shows that combined analyses of bioinformatic and experimental data enable to suggest the existence of previously unknown biochemical phenomena.

Further points raised are the involvement of nucleotide dipole moments in the interactions between the nucleotide and the gamma polymerase. Analyses at this point do not yield much information beyond the fact that such effects occur. More functional hypotheses could help in this respect.

The data on mutation patterns from *Homo sapiens* do not enable to establish whether mutations cumulate during singlestrandedness due to transcription or replication. It is probable that the ratio between these two types of events that open double stranded DNA, changes with the longevity of an individual/species, where greater lifespan increases the transcription component [11]. It is probable that analyses similar to those done here based on ample mitochondrial sequence data available for other mammal species with shorter lifespans could help in this respect. Comparing results from different species would probably be fruitful. In addition, these analyses could preliminarily reveal whether misincorporation patterns by the gamma polymerases of the different species differ. This could be an exciting line of research, that could potentially link differences in mutation patterns with differences in the gamma polymerases from these species. Such analyses could yield a workable model for the efficiency and fidelity of gamma polymerase in relation to its detailed structure. It is notable that much information necessary for such analyses is already available online and only awaits the interest of enthousiastic students of molecular biology.

Nearest neighbour effects as detected for mitochondrial mutation patterns confirm what is known from previous studies on nuclear chromosomes. They also show that the 5' G effect on mutation rates is more complex, as it affects differently different types of mutations. Unfortunately, nearest neighbour analyses require samples that are not compatible with the data at hand, so that its analysis in combination with other factors could not be done. The fact that nearest neighbour effects tend to increase (though marginally so), with the thermodynamic stability of the DNA duplex where these neighbouring effects occur, is in itself compatible with dipole moment effects as the causes for the nearest neighbour effects on mutations because nucleotide dipole moments predict duplex thermodynamic stabilities [37]. It is possible that the direct cause for this is thermodynamic stability, through the fact that regions forming stable duplexes are more able to tolerate a misinserted nucleotide. But the association between nearest neighbour effects and stability is weak, indicating that another, associated factor is at work. Possibly, it is the electrostatic effect of nucleotide dipole moments of neighbouring nucleotides on the fidelity of the gamma polymerase that causes these effects. Such effects are particularly probable, considering that the gamma polymerase active site includes two charged residues, and that nucleotide processing seems to depend to some extent on the nucleotide's dipole moments. Hence nearest neighbour effects could be due to interferences between the electrostatic fields of the active site, the incorporated nucleotide, and the nearest neighbours, especially when these nearest neighbours have high dipole moment.

Beyond effects of nucleotide dipole moments on incorporation rates, results suggest natural selection decreasing nucleotide misincorporations with high impact on protein structure. These are encouraging results that could yield further insights if similar analyses are applied to different types of polymerases.

Variation in mutation patterns for genes with different locations along singlestrandedness gradients might have explanations that differ from the ones suggested. The genome structure might be designed so that genes that cannot afford, from a functional point of view, large mutation rates, are located so as to endure little singlestrandedness. It is important to remember that this factor might interact with the results presented. It is also important to remember that natural selection probably affects observed mutation frequencies in ways not accounted for by presented analyses. This effect might be weaker in genes located far from replication origins, and hence probably more able to tolerate mutations. Hence observed mutations would in these cases much more reflect the original processes, not confounded by effects of natural selection due to gene function.

A further point relates to the patterns observed in Figure 5, where gene length seems to affect the mutation pattern. A possible factor here is the capacity of longer sequences to form more secondary structures by self-hybridization. Considering that secondary structure protects against mutations due to singlestrandedness, this factor could hence indirectly affect mutation patterns, especially in longer genes, assuming that in some ways, genes are replicated as functional units, a possibility that cannot be ruled out *a priori*, especially if secondary structure formation is designed to involve a gene as a unit, for example in the mRNA [38]. It is also possible that secondary structures affect the function of the gamma polymerase, causing differences in misincorporation patterns between regions forming more or less secondary structures, as previous analyses possibly indicated [13, 14].

An important point to stress here is that the data that are available at this point do not limit our capacities to analyses, along multiple dimensions, the various factors that cause mutation patterns, and understand their details in relation to these factors. The computational power and statistical tools are also not limiting and close to adequate. The limiting factor is the time invested by the adequately skilled manpower, or more correctly, the financial investment to support such activity based mainly on analysing valuable molecular data of different types.

15. Conclusions

Combined analyses of comparative sequence data and experimentally determined gamma polymerase misincorporation data, together with models for substitutions based on nucleotide dipole moments and models for substitution impacts on protein structure reveal that observed human mitochondrial protein coding gene mutation patterns are affected in decreasing order of importance by gamma polymerase misincorporation rates, selection against non-conservative amino acid replacements, and gradients in singlestrandedness during replication and transcription. Gamma polymerase misincorporation rates are selected to optimize effects of substitutions on non-conservative amino acid replacements, and favour nucleotides with low dipole moments, suggesting hydrophobic bias in nucleotide misbinding. Further analyses confirm this: the hydrophilic active site of the gamma polymerase handles faster nucleotides with high dipole moment and mishandles more often those with low dipole moment, suggesting that process accuracy limits its rate. The wealth of results confirms known and expected patterns, and expands beyond them, revealing selection on polymerase fidelity, and spontaneous tendencies during single stranded DNA states for all substitutions, not only those previously known to react to singlestrandedness.

Author details

Hervé Seligmann*

Address all correspondence to: podarcissicula@gmail.com

The National Collections of Natural History at the Hebrew University of Jerusalem, Israel

References

[1] Ingman, M., & Gyllenstein, U. (2006). mtDB: Human mitochondrial genome database, a resource for population genetics and medical sciences. Nuc Acids Res, 34, D749-D751.

[2] Ruiz-Pesini, E., Lott, M. T., Procaccio, V., Poole, J., Brandon, M. C., Mishmar, D., Yi, C., Kreuziger, J., Baldi, P., & Wallace, D. C. (2007). An enhanced MITOMAP with a global mtDNA mutational phylogeny. Nuc Acids Res, 35(Database issue: D823-D828).

[3] Kaguni, L. S. (2004). DNA polymerase gamma, the mitochondrial replicase. Ann Rev Biochem, 73, 293-320.

[4] Seligmann, H. (2006). Error propagation across levels of organization: From chemical stability of ribosomal RNA to developmental stability. J Theor Biol, 242, 69-80.

[5] Li, W. H., Wu, C. I., & Luo, C. C. (1984). Nonrandomness of point mutation as reflected in nucleotide substitutions in pseudogenes and its evolutionary implications. J Mol Evol, 21, 58-71.

[6] Bergmann, E. D., & Weiler-Feilchenfeld, H. (1973). The dipole moments of purines and pyrimidines. Chapter 1. In: Duchesne J (ed) Physico-chemical properties of nucleic acids, I., Academic Press, London.

[7] Frederico, L. A., Kunkel, T. A., & Shaw, B. R. (1990). A sensitive genetic assay for the detection of cytosine deamination: determination of rate constants and the activation energy. Biochem, 29, 2532-2537.

[8] Frederico, L. A., Kunkel, T. A., & Shaw, B. R. (1993). Cytosine deamination in missmatched base-pairs. Biochem, 32, 6523-6530.

[9] Francino, M. P., Chao, L., Riley, M. A., & Ochman, H. (1996). Asymmetries generated by transcription-coupled repair in enterobacterial genes. Science, 272, 107-109.

[10] Francino, M. P., & Ochman, H. (1993). Deamination as the basis of strandasymmetric evolution in transcribed Escherichia coli sequences. Mol Biol Evol, 18, 1147-1150.

[11] Seligmann, H. (2011). Mutation patterns due to converging mitochondrial replication and transcription increase lifespan, and cause growth rate-longevity tradeoffs. *In: Seligmann H. (ed.) DNA Replication-Current Advances*, Rijeka, InTech, Chapter 6, 151-180.

[12] Seligmann, H. (2012). Coding constraints modulate chemically spontaneous mutational replication gradients in mitochondrial genomes. *Curr Genomic*, 13, 37-54.

[13] Krishnan, N. M., Seligmann, H., Raina, S. Z., & Pollock, D. D. (2004). Detecting gradients of asymmetry in site-specific substitutions in mitochondrial genomes. *DNA & Cell Biol*, 23, 707-714.

[14] Krishnan, N. M., Seligmann, H., Raina, S. Z., & Pollock, D. D. Phylogenetic analysis of site-specific perturbations in asymmetric mutation gradients. *In: A. Gramada, and P.E. Bourne (eds.) Currents in Computational Molecular Biology*, ACM Press, San Diego, CA, 266-267.

[15] Seligmann, H., Krishnan, N. M., & Rao, B. J. (2006). Possible multiple origins of replication in primate mitochondria: alternative role of tRNA sequences. *J Theor Biol*, 241, 321-332.

[16] Madariaga, S. T., Contreras, J. G., & Seguel, C. G. (2005). Interaction energies in non Watson-Crick pairs: an ab initio study of G U and U U pairs. *J Chil Chem Soc*, 50, 435-438.

[17] Gojobori, T., Li, W. H., & Graur, D. (1982). Patterns of nucleotide substitution in pseudogenes and functional genes. *J Mol Evol*, 18, 360-369.

[18] Lee, H. R., & Johnson, K. A. (2006). Fidelity of the human mitochondrial DNA polymerase. *J Biol Chem*, 281, 36236-36240.

[19] Lee, Y. S., Kennedy, W. D., & Yin, Y. W. (2009). Structural insight into processive human mitochondrial DNA synthesis and disease-related polymerase mutations. *Cell*, 139, 312-324.

[20] Grantham, R. (1974). Amino acid difference formula to help explain protein evolution. *Science*, 185, 862-864.

[21] Seligmann, H. (2008). Hybridization between mitochondrial heavy strand tDNA and expressed light strand tRNA modulates the function of heavy strand tDNA as light strand replication origin. *J Mol Biol*, 379, 188-199.

[22] Seligmann, H., Krishnan, N. M., & Rao, B. J. (2006). Mitochondrial tRNA sequences as unusual replication origins: pathogenic implications for Homo sapiens. *J Theor Biol*, 243, 375-385.

[23] Seligmann, H. (2010). Mitochondrial tRNAs as light strand replication origins: similarity between anticodon loops and the loop of the light strand replication origin predicts initiation of DNA replication. *Biosystems*, 99, 85-93.

[24] Tanaka, M., & Ozawa, T. (1994). Strand asymmetry in human mitochondrial mutations. *Genomics*, 22, 327-335.

[25] Seligmann, H. (2010). Positive correlations between molecular and morphological rates of evolution. *J Theor Biol*, 264, 799-807.

[26] Krawczak, M., Ball, E. V., & Cooper, D. N. (1998). Neighboring-nucleotide effects on the rates of germ-line single-base-pair substitution in human genes. *Am. J. Hum. Genet.*, 63, 474-488.

[27] Zhao, Z., & Boerwinkle, E. (2002). Neighboring-nucleotide effects on single nucleotide polymorphisms: a study of 2.6 million polymorphisms across the human genome. *Genome Res*, 12, 1679-1686.

[28] Zhanga, F., & Zhao, Z. (2004). The influence of neighboring-nucleotide composition on single nucleotide polymorphisms (SNPs) in the mouse genome and its comparison with human SNPs. *Genomics*, 84, 786-796.

[29] Zhao, H., Li, Q.-Z., Zeng, C.-Q., Yang, H.-M., & Yu, J. (2005). Neighboring-nucleotide effects on the mutation patterns of the rice genome. *Geno. Prot. Bioinfo.*, 3(3).

[30] Zhang, W., Bouffard, G. G., Wallace, S. S., & Bond, J. P. (2007). NISC Comparative Sequencing Program. Estimation of DNA sequence context-dependent mutation rates using primate genomic gequences. *J Mol Evol*, 65, 207-214.

[31] Suzuki, Y., Gojobori, T., & Kumar, S. (2009). Methods for incorporating the hypermutability of CpG dinucleotides in detecting natural selection operating at the amino acid sequence level. *Mol Biol Evol*, 26, 2275-2284.

[32] Cheunga, L. W. T., Leeb, Y. F., Ngb, T. W., Chingb, W. K., Khooc, U. S., Ngd, M. K. P., & Wong, A. S. T. (2007). CpG/CpNpG motifs in the coding region are preferred sites for mutagenesis in the breast cancer susceptibility genes. *FEBS Letters*, 681, 4668-4674.

[33] Antonarakis, S. E. (2006). *CpG Dinucleotides and Human Disorders. eLS.*

[34] Misawa, K. (2011). A codon substitution model that incorporates the effect of the GC contents, the gene density and the density of CpG islands of human chromosomes. *BMC Genomics*, 12, 397.

[35] Chen-L, C., Rappailles, A., Duquenne, L., Huvet, M., Guilbaud, G., Farinelli, L., Audit, B., d'Aubenton-Carafa, Y., Arneodo, A., Hyrien, O., & Thermes, C. (2010). Impact of replication timing on non-CpG and CpG substitution rates in mammalian genomes. *Genome Res*, 20, 447-457.

[36] Kasiviswanathan, R., & Copeland, W. C. (2011). Ribonucleotid discrimination and reverse transcription by the human mitochondrial DNA polymerase. *J Biol Chem*, in press.

[37] Seligmann, H., & Amzallag, G. N. (2002). Chemical interactions between amino acid and RNA: multiplicity of the levels of specificity explains origin of the genetic code. *Naturwissenschaften*, 89, 542-551.

[38] Krishnan, N. M., Seligmann, H., & Rao, B. J. (2008). Relationship between mRNA secondary structure and sequence variability in chloroplast genes: possible life history implications. *BMC Genomics*, 9, 48.

Chromatin and Epigenetic Influences on DNA Replication

Chapter 3

The Mechanisms of Epigenetic Modifications During DNA Replication

Takeo Kubota, Kunio Miyake and Takae Hirasawa

Additional information is available at the end of the chapter

1. Introduction

At the DNA replication step during cell division, not only fundamental information (i.e. nucleotide sequence) but also superficial information (i.e. "epigenetic" modifications) is faithfully reproduced on the newly synthesized DNA sequence. The faithful maintenance of the epigenetic pattern, which determines the gene-expression pattern of the cell, safeguards the maintenance of cell identity.

The term "epigenetics" was first used to describe "the causal interactions between genes and their products, which bring the phenotype into being" [1], and this definition initially referred to the role of the epigenetics in embryonic development, in which cells develop distinct identities despite having the same genetic information. However, today epigenetics refers to "the study of heritable changes in gene expression that occur independent of changes in the primary DNA sequence" [2]. This definition is now associated with in a wide variety of biological processes, such as genomic imprinting [3,4], inactivation of the X chromosome [5], embryogenesis [6], tissue differentiation [7], and carcinogenesis [8].

Epigenetic chemical modifications, such as DNA methylation and histone modifications, are known to be faithfully duplicated in each cell cycle and subsequently the chromatin structures are propagated through DNA replication [9]; however, little is known about how the chromatin structure is maintained during or reformed after DNA replication. Furthermore, several lines of recent evidence suggested that the superficial information on the DNA strand is more susceptible to change by environmental stress than the DNA strand itself. Therefore, for a better understanding of the DNA replication process, it is highly important and desirable, for biologists in general and molecular biology in particular, to learn about the epigenetic mechanisms.

In this chapter, we introduce the current understanding of the DNA methylation mechanism, 5-hydroxycytosine (the sixth base), histone modifications, and their significance in congenital and acquired diseases, and also discuss to which direction this field ought to proceed in the future.

2. DNA methylation during DNA replication

Not all genes are necessarily expressed in every cell of the organism. Most of these genes and genetic regions are programmed to remain repressed, which defines the identity of each cell. Epigenetic modifications are molecular mechanisms that can preserve the inactive state by regenerating a repressive chromatin structure on the "unnecessary genes and genomic regions" following each round of DNA replication in the cell. DNA methylation is one of the fundamental mechanisms known to be involved in this maintenance process [10].

Maintenance of such methylation pattern in DNA during replication is mediated by DNA nucleotide methyltransferase 1 (DNMT1) [11], which methylates newly synthesized CpG sequences, depending on the methylation status of the template strand (Fig. 1). A bridging protein, known as UHRF1 (ubiquitin-like, containing PHD and RING finger domains 1), that interacts with DNMT1 and hemimethylated CpG is required to maintain the hemimethylated CpG dinucleotides pattern at the DNA fork [12,13].

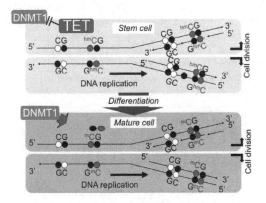

Figure 1. Maintenance of DNA methylation pattern with 5-hydroxymethylcytosine (upper) and 5-methylcytosine (lower) during DNA replication and cell division. 5-hydroxymethylcytosine is relatively abundantly found in embryonic stem (ES) cells and its level decreases during development due to the declining levels of *TET* expression. Cytosine, 5-hydroxymethylcytosine and 5-methylcytosine are shown in white, blue and red circles, respectively [95].

The chromatin structure, modified by DNA methylation, is not stable, but it undergoes a wave of disruption and reassembly during DNA replication. These changes in the chromatin structures influence the dynamics of DNA replication by regulating the selection of replication origin sites and their initiation timings. Interestingly, active gene promoters are often found at these active replication origin sites. Thus, the coordination of replication and transcription is an important mechanism for the establishment and inheritance of differential gene expression patterns during cellular differentiation [2].

DNA methylation status is also involved in determining the chromosomal replication timing. Hypomethylation is associated with late-replication and late-replicating genomic regions are gradually demethylated with cell divisions, whereas DNA methylation of early-replicating regions is maintained during DNA replication [14]. Moreover, DNA replication in early S phase gets automatically repackaged with acetylated histones, whereas the regions that replicate late in S phase assemble nucleosomes containing deacetylated histones [15].

So far several DNA nucleotide methyltransferases (DNMTs), which includes DNMT1, DNMT2, DNMT3A, DNMT3B, DNMT3L, have been found in mammals, all of which contain a methyltransferase catalytic domain. Of these, DNMT1 is the most abundant DNMT in differentiated cells; it has a preference for hemi-methylated DNA, and acts as a 'maintenance methylase', which allows it to efficiently methylate the hemi-methylated sites that are generated during DNA replication. Thus, the CpG methylation pattern is maintained in the genome after DNA replication [16]. Until recently, the biochemical and functional properties of DNMT2 remained unknown. However, the DNMT2 is now known to act as an RNA methylatransferase and the DNMT2-mediated methylation protects tRNAs against ribonuclease cleavage in drosophila [17].

DNMT3A and DNMT3B are expressed at high levels in mouse embryonic stem (ES) cells and at lower levels in differentiated cells. They act as 'de novo methylases', which catalyze the transfer of methyl groups to naked DNA, and are responsible for establishing the pattern of methylation during embryonic development [18]. Recent evidence suggests that besides playing the role as 'de novo methylases' DNMT3A and DNMT3B may also act as 'methylation completer' and 'methylation error corrector' - by completing the methylation process and correcting errors, respectively, left by DNMT1 - at least at highly methylated DNA regions, such as imprinted regions and repetitive elements [19].

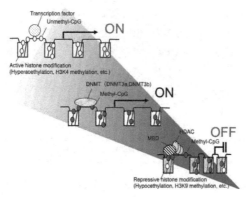

Figure 2. Epigenetic gene regulation based on DNA methylation, histone acetylation and histone methylation, induced by proteins including nucleotide methyltransferases (DNMTs), methyl-CpG binding domain (MBD) proteins, and histone deacetylases (HDACs).

Once a certain site is methylated, it could then act as a candidate region where the silent chromatin is established. For this purpose, the methylated site first recruits methyl-CpG binding domain (MBD) proteins; the MBD proteins subsequently recruit histone deacetylases, histone modification proteins. In other words, MBDs, which form bridges between the methylation site and other associated proteins, are the key proteins in epigenetic regulation (Fig. 2).

So far five MBD proteins, each containing a methyl-CpG binding domain, have been reported. Among these MBD proteins, MBD1 is unique because it is capable of repressing transcription from both methylated and unmethylated promoters [20].

MBD1 associates with chromatin modifiers such as the Suv39h1-HP1 complex, and enhances DNA methylation-mediated transcriptional repression [21]. MBD1 also associates with the H3K9 methyltransferase SETDB1 [22]. During S phase, the chromatin assembly factor CAF1 recruits the MBD1-SETDB1 complex to chromatin to establish new H3K9 methylation. On the other hand, the removal of DNA methylation disrupts the formation of MBD1-SETDB1-CAF1 complex, which results in the loss of H3K9 methylation at the formerly methylated site [23].

MBD2 protein shares extensive sequence homology with MBD3. MBD2 binds to methylated CpGs and confer DNA methylation-mediated transcriptional silencing through its association with HDAC1 and HDAC2 in the NuRD chromatin remodeling complex [23]. Although *Mbd2*-null mice develops normally and remains viable and fertile [24], lack of Mbd2 affects immunological systems by inducing ectopic *IL-4* expression in undifferentiated helper T cells [25]. Lack of Mbd2 also influences X-chromosome inactivation by inducing ectopic *Xist* expression in the active X chromosome [26].

MBD3, like MBD2, is an essential subunit of the NuRD complex. It has been suggested that MBD2 and MBD3 associate with the NuRD in a mutually exclusive way, thereby forming two distinct complexes [23]. Although there is a great sequence similarity between MBD2 and MBD3, the two proteins do not perform redundant functions during early development. In contrast to *Mbd2*-mull mice which displayed a mild phenotype, *MBD3*-null embryos die on day 8.5, by failing to shut down the expression of undifferentiated cell markers such as *Oct4* and *Nanog* [27].

MBD4 is a thymine glycosylase, which acts as a DNA repair protein and targets the sites of cytosine deamination. Spontaneous hydrolytic deamination of 5mC leads to 5mCpG-TpG transitions, whereas that of non-methylated CpG leads it to UpG, and MBD4 is able to excise and repair both 'mutated' nucleotides [28]. Consistent with this observation, *Mbd4*-null mice exhibit a two to three times higher number of 5mCpG-TpG transitions, indicating that Mbd4 indeed acts to reduce the 5mCpG-TpG mutation rate [29]. More importantly, when crossed with mice carrying a germline mutation in the *Apc* (adenomatous polyposis coli) gene, *Mbd4*-null mice show accelerated tumor formation [29]. In fact, mutations in *MBD4* have been reported in various human carcinomas [30].

MeCP2 is the first MBD to be cloned [31]. As of now, MeCP2 is known to be a multifunctional nuclear protein, which is known to be involved transcriptional repression, activation of transcription, nuclear organization, and splicing [32,33]. Besides acting as a transcriptional repressor like other MBDs, MeCP2 also acts as a splice regulator, by interacting with YB-1, a component of messenger ribonucleoprotein particles, in brain nuclear extracts [34]. Indeed, microarray splicing analysis of cerebral cortex mRNA isolated from *Mecp2*-mutant mice showed a number of aberrantly spliced genes [23]. Furthermore, MeCP2 deficiency activates L1 retrotransposition in neurons, which is possibly associated with the genomic diversity of brains [35]. Therefore, it is interesting that there exist several links between MBD-mediated repression, RNA processing and DNA-sequence diversity. It is also intriguing to find a link between epigenetic modification and its suppressive power on genetic diversity since, in addition to MeCP2, a histone modification enzyme (H3K9 methyltransferase, *ESET*) also contributes to silence retrovirus-like elements in the mammalian genome [36].

3. 5-hydroxymethylcytosine - the sixth base in mammalian DNA

The 5-methylcytosine (5mC) has been recognized as "the fifth base". However, early work suggested the existence of a sixth base, 5-hydroxymethylcytosine (5hmC) (Fig. 1). 5hmC was first reported in T-even bacteriophages [37] and later in mammalian cells [38]. However, the reported finding, which claimed that this modified base accounted for ~15% of total cytosines in DNA extracted from the brains of adult rats, mice and frogs, could not be reproduced [39]. The topic received only little attention for the next 30 years until 2009, when work from two research teams brought it back to life [40,41]. Actually, it was found that 5hmC accounts for 0.6%, 0.2%, 0.03% of total nucleotides in Purkinje cells, granule cells, and mouse ES cells, respectively [40,41].

The presence of 5hmC in the mammalian genome depends on pre-existing 5mC, because 5hmC is converted from 5mC with the help of TET proteins, which utilize molecular oxygen to incorporate a hydroxyl group to 5mC. TET is named after Ten-Eleven Translocation (translocation between chromosomes 10 and 11) because it is initially found as a fusion protein partner of mixed-lineage leukemia gene (*MLL*) in acute myeloid leukemia (AML) patients carrying a t(10;11)(q22;q23) translocation [42,43]. The findings that ectopic expression of TET1 in HEK293 cells lacking TET1 led to reduced levels of 5mC and increased levels of 5hmC, and that the levels of 5hmC decreased upon RNAi-mediated depletion of TET1 in ES cells indicate that TET1 is able to catalyze the conversion of 5mC and 5hmC in cultured cells [41]. Also, it has been demonstrated that TET1 is capable of acting not only on fully-methylated DNA strands but also on hemi-methylated DNA strands [41]. Furthermore, not only TET1 but also other TET proteins (TET2 and TET3) is capable of converting 5mC to 5hmC [44].

In terms of gene regulation, the significance of this 5hmC modification is similar to that of non-methylated cytosine. In other words, the 5hmC modification is associated with transcriptional activity, which is different from the 5mC modification that is associated with transcriptional repression [16]. It has recently demonstrated that TET1-binding to the promoter region (presumably 5hmC modification at this site) induces the expression of *Nanog* in ES cells and that downregulation of *Nanog* via *TET1* knockdown induces DNA methylation in the promoter region [44]. These findings indicate that the TET1 driven 5hmC modification contribute to maintenance of the nature of un-differentiation and pluripotency of ES cells, and support a working model by which TET1 and DNMTs coordinately regulate Nanog expression.

In ES cells, high levels of TET1 block the access of DNMTs for maintained Nanog expression. On the other hand, when TET1 is downregulated in ES cells by *in vitro* differentiation, DNMTs methylate the *Nanog* promoter, which leads to the downregulation of Nanog expression and loss of ES cell identity (Fig. 1) [44]. This hypothesis is supported by a recent finding in which the chromosomes containing 5hmC are gradually reduced during the development of preimplantation embryos [45]. However, another study showed that the 5hmC level in the mouse cerebellum during development increases from 0.1% of total nucleotides at postnatal day 7 to 0.4% of total nucleotides in the adult mouse [46].

As described above, TET1 was initially identified through a rare translocation case with leukemia [42,43]. Later studies have demonstrated that deletion and mutations in *TET1*, *TET2* and *TET3* are associated with myeloid malignancies [47]. In fact, mutations found in *TET2* in myeloid cancers have been shown to impair hydroxylation of 5mC [48].

While our knowledge about 5hmC is rapidly growing, currently there is no reliable methodology available that would provide information on 5hmC at single-base-pair resolution. Although a 5hmC antibody is available for chromatin Immunoprecipitation, this method only provides some coarse information (i.e. detects presence of 5hmC but not that of 5mC in chromatin). A more sensitive method has been developed for 5hmC by capillary eletrophoresis, but this is not the one at sinlge-base-pair resolution [49]. Another method (namely, bisulfite sequencing) has proven to be a powerful tool for providing information on the methylation status at single-base-pair resolution. However, it too fails to discriminate between 5mC and 5hmC. Thus, if the bisulfite-treated DNA is used as a template for PCR analysis, cytosine will be read as thymine, whereas both 5mC and 5hmC will be read as cytosine [16]. Therefore, it is important to develop a methodology that can distinguish between 5mC and 5hmC at single-base-pair resolution in order to achieve complete understanding of the active demethylation mechanism, because TET protein-mediated 5mC oxidation may contribute to dynamic changes in global or locus-specific 5mC levels by promoting active DNA demethylation [50].

4. Histone modifications and DNA methylation during replication

DNA methylation and histone modifications not only occur separately, but they also work hand-in-hand at multiple levels to determine expression status, chromatin organization and cellular identity, and they are co-ordinately maintained through mitotic cell division, allowing for the transmission of parental DNA and for the histone modifications to be copied to newly replicated chromatin [51,52].

Lande-Diner et al. recently developed a *DNMT1*- knockout cell line and demonstrated that an unmethylated state, caused by the lack of DNMT1, induced deacetylation of histones H3 and H4, resulting in transcriptional activation in many genes [10]. This observation clearly indicates that DNA methylation is associated with histone deacetylation. However, this group also demonstrated that in several other genes the unmethylated state, caused by lack of DNMT, did not induce histone H3 and H4 deacetylation, resulting in transcriptional repression. In addition, late replication in S phase was observed at these loci, suggesting that the replication timing may be independent of DNA methylation [10]. Rather, histone acetylation is associated in controlling the replication timing [53].

DNA methylation is not only correlated to histone 'acetylation', but also associated with histone 'lysine methylation'. Genome-wide DNA methylation profiles suggest that DNA methylation is associated with the absence H3K4 methylation and the presence of H3K9 and H3K27 methylation [54].

In fact, DNA methylation induces histone H3K9 methylation through an MBD, thereby establishing a repressive chromatin state [55]. SETDB1, a H3K9 trimethylation (H3K9me3) methyltransferase, contains a putative MBD domain with two conserved DNA-interacting arginine residues, which are also present in the MBD domains of MBD1 and MeCP2 and are known to make direct contact with the DNA in the structures of MBD1-DNA and MeCP2-DNA complexes [56,57]. This result suggests that SETDB1 acts as an H3K9me3 'writer' in corporation with DNA methylation 'reader'. Likewise, SUV39H1/2, another H3K9me3 'writer', interacts with HP1, the H3K9me3 'reader' to create a repressed status in their recruited genomic region [58]. These are the mechanisms for propagating and maintaining repressive chromatin marks on both DNA and histones during DNA replication.

A histone methyltransferase, in turn, can direct DNA methylation to specific genomic targets by recruiting DNMTs to stably silence genes [59]; accordingly, disruption of the histone lysine methyltransferase gene with specificity for H3K4 (*MLL*) in mice not only induces the loss of H3K4 methylation but also induces de novo DNA methylation at several gene promoters [60,61]. In another study, it was shown that the lack of histone H3K9 methyltransferase induced demethylation at the imprinting center in *SNPRN* locus on the maternal chromosome, whereas the lack of DNMT1 failed to induce demethylation of histone H3K9, indicating that the modification order at this locus is histone modification followed by DNA methylation [62]. Taken together, histone methylation marks play important roles in predicting the methylation status of the genome [63].

Whereas DNMT1 is stabilized by a histone demethylase (HDM) to maintain DNA methylation [64], DNMTs can direct the local status of histone methylation patterns, recruiting MBDs and HDACs to achieve gene silencing and chromatin condensation [65,66]. Recently, DNMT3L has been shown to act as a sensor for H3K4 methylation. Thus, when methylation is absent, DNMT3L induces *de novo* DNA methylation by docking DNMT3A to the nucleosome, which is one of mechanisms by which methylated regions are newly created during the replication step [67].

The interplay of these modifications creates an epigenetic landscape that regulates the way the mammalian genome expresses itself in different cell types, developmental stages and disease states. The distinct patterns of these epigenetic modifications present in different cellular states serve as a guardian of cellular identity [2]. Whereas it is well accepted that DNA methylation patterns are replicated in a semi-conservative fashion during cell division via the mechanisms discussed earlier, how histone modification patterns are similarly replicated remains to be elucidated.

5. Abnormalities in epigenetic mechanism and their possible inheritance

Thanks to identification of molecules that contribute to epigenetic gene regulation, we now know how that abnormalities in these molecules cause a number of congenital diseases.

The first group of diseases with abnormal epigenetic mechanism is genomic imprinting diseases [3]. Genomic imprinting is a mechanism in which only one of the two parental alleles is expressed in a gene. For example, in the case of *SNRPN* gene, the paternal allele of the *SNRPN* gene is expressed, whereas the maternal allele is suppressed by DNA methylation in normal individuals, and abnormal suppression of the normally expressing paternal allele causes a congenital obesity disease, known as Prader-Willi syndrome [4]. In the case of *UBE3A* gene, which locates adjacent to the *SNRPN* gene, the maternal allele is expressed, whereas the paternal allele is suppressed in neurons [68]; abnormal suppression of the expressing maternal allele causes a congenital epileptic disease, known as Angelman syndrome [69].

X-chromosome inactivation is another epigenetic mechanism in which only one of the two X chromosomes is activated and the other X chromosome is inactivated in females [5]. Females with aberrant X-inactivation (i.e. both two X chromosomes are activated) are thought to be embryonic lethal, since somatic clones with aberrant X-inactivation are aborted [70].

Abnormal functioning of the proteins related to epigenetic regulation also causes diseases. For example, mutations in the *DNMT3B* gene, which lead to hypomethylation at the paracentromeric chromosomal regions, cause the immunodeficiency- centromeric instability- facial anomalies (ICF) syndrome, which is characterized by immunodeficiency, centromere instability, facial abnormalities, and mild mental retardation (Fig. 3A) [71-73]. On the other hand, over-expression of *DNMTs* is associated with hypermethylation found in colorectal, breast, and hepatocellular carcinomas (Fig. 3C) [74-76]. Another example is Rett syndrome caused by *MECP2* mutations, which is characterized by seizures, ataxic gait, language dysfunction and autistic behavior [77,78]. In this disease, *MECP2* mutations induce abnormal regulation of a subset of neuronal genes [79,80] (Fig. 3B).

Besides these "DNA methylation diseases" caused by mutations in DNA methylation-related enzymes and proteins, "histone modification diseases" caused by mutations in histone modification-related enzymes have recently been reported. For example, Say-Barber-Bieseker-Young-Simpson syndrome is caused by mutations in the histone acetyltransferase gene, *KAT6B*, which is a multiple anomaly syndrome characterized by, an immobile mask-like face, abnormal narrowing of palpebral fissures (short eyelid), anomalies of the spine, ribs and pelvis, renal cysts, hydronephrosis, agenesis of the corpus callosum, and severe intellectual disability [81]. Another example is Kleefstra syndrome caused by deletion or mutation in the histone H3K9 methyltransferase gene, *EHMT1*, which is characterized by childhood hypotonia, distinctive facial features, and intellectual disability with severe expressive speech delay [82].

Recently, it has been shown that short-term environmental stress could also cause aberrant epigenetic status associated with various diseases. Thus, aberrant epigenetic mechanism can not only cause congenital diseases, but can also cause acquired diseases. For example, short-term mental stress after birth, in which the mother is separated from the offspring, causes DNA hypermethylation in the promoter of the glucocorticoid receptor (*GR*) gene in the rat brain, resulting in persistent abnormal behavior [83]. Malnutrition in the fatal period is also known to induce DNA hypomethylation in the promoter of the peroxisome proliferator-activated receptor alpha (*PPARα*) gene, a so-called "thrifty gene", in the liver, which may be associated with the developmental basis of adult diseases (i.e. obesity and diabetes mellitus) [84,85] (Fig. 3D). This hypomethylation event has later been confirmed in human individuals who suffered prenatal malnutrition during the period of famine [86,87].

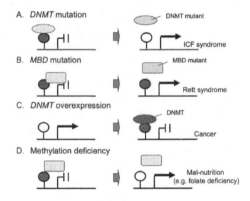

Figure 3. Abnormal epigenetic patterns found in human congenital and acquired diseases.

Several lines of evidence suggested that acquired DNA methylation changes described above are transmitted to the next generation. Epigenetic marks allow the transmission of gene activity states from one cell to its daughter cells. Initially, it was assumed that epigenetic marks were completely erased and re-established in each generation. However, recent studies using several model organisms indicate that the erasing process is incomplete at some loci and so the epigenetic changes acquired in one generation are inherited by the next generation.

For example, it has been shown in mice that the mental stress caused due to maternal separation in offspring not only changes the DNA methylation status in the first generation but also in the next generation through changes in the sperms of the first generation in mice [88]. Moreover, the observed changes in the DNA methylation status altered the expression level of corticotrophin releasing actor receptor 2 (*Crfr2*) in the brains of next generation mice, which could be associated with their abnormal behavior [88].

This phenomenon is termed "transgenerational epigenetic inheritance", which is expected to provide a biological proof for the apparent heritability of acquired characteristics [89-91].

6. Concluding remarks

One of the major differences between DNA sequence and epigenetic modifications is tissue specificity. Epigenetic modifications vary according to the tissue type, which consequently allows generating tissue-specific expression patterns. However, how determines the epigenetic modification (epigenomic) pattern in each tissue type is not fully understood.

Thus, it is essential to categorize epigenomic patterns in each human tissue at the nucleotide resolution [92,93]. In fact, the NIH Roadmap Epigenomics Program under the US National Center for Biotechnology Information (NCBI) and the International Human Epigenome Consortium (IHEC) have initiated the large-scale epigenomic mapping studies in order to generate epigenome maps for each human cell type for this purpose [94].

Understanding the human epigenome will be fundamental to the study of congenital and acquired diseases, and will also be invaluable for analyzing the linkage between birth defects and environmental factors. However, biological studies to understand the epigenome are in their initial phase. Further studies are necessary to elucidate the molecular mechanism by which the epigenome pattern in each cell type differs, epigenomic patterns are altered by environmental factors, and process of inheriting the epigenomic pattern from the previous generation could be avoided. The authors expect that these molecular mechanisms would hopefully be discovered by the "next generation" of researchers.

Acknowledgements

The research described in this article was partially supported by the Ministry of Education, Science, Sports and Culture (MEXT), grants-in-aid (KAKENHI) for Scientific Research (B) (23390272) to TK, grants-in-aid for Exploratory Research (23659519) to TK, grants-in-aid for Young Scientists (B) (23791156) to KM, and grants-in-aid for Scientific Research (C) (23591491) to TH. The authors thank NAI inc. for critical review.

Author details

Takeo Kubota*, Kunio Miyake and Takae Hirasawa

*Address all correspondence to: takeot@yamanashi.ac.jp

Department of Epigenetic Medicine, Faculty of Medicine, University of Yamanashi, Japan

References

[1] Waddington, C. H. (1942). Epigenotype. *Endeavour*, 1, 18-20.

[2] Sharma, S., Kelly, T. K., & Jones, P. A. (2010). Epigenetics in cancer. *Carcinogenesis*, 31(1), 27-36.

[3] Kubota, T., Saitoh, S., Matsumoto, T., Narahara, K., Fukushima, Y., Jinno, Y., & Niikawa, N. (1994). Excess functional copy of allele at chromosomal region 11-15 may cause Wiedemann-Beckwith (EMG) syndrome. *American Journal of Medical Genetics*, 49(4), 378-83.

[4] Kubota, T., Das, S., Christian, S. L., Baylin, S. B., Herman, J. G., & Ledbetter, D. H. (1997). Methylation-specific PCR simplifies imprinting analysis. *Nature Genetics*, 16(1), 16-7.

[5] Kubota, T., Nonoyama, S., Tonoki, H., Masuno, M., Imaizumi, K., Kojima, M., Wakui, K., Shimadzu, M., & Fukushima, Y. (1999). A new assay for the analysis of X-chromosome inactivation based on methylation-specific PCR. *Human Genetics*, 104(1), 49-55.

[6] Sasaki, H., & Matsui, Y. (2008). Epigenetic events in mammalian germ-cell development: reprogramming and beyond. *Nature Reviews Genetics*, 9(2), 129-40.

[7] Sakashita, K., Koike, K., Kinoshita, T., Shiohara, M., Kamijo, T., Taniguchi, S., & Kubota, T. (2001). Dynamic DNA methylation change in the CpG island region of p15 during human myeloid development. *J Clin Invest. Journal of Clinical Investigation*, 108(8), 1195-204.

[8] Ushijima, T. (2005). Detection and interpretation of altered methylation patterns in cancer cells. *Nature Reviews Cancer*, 5(3), 223-31.

[9] Liu, Q., & Gong, Z. (2011). The coupling of epigenome replication with DNA replication. *Current Opinion in Plant Biology*, 14(2), 187-94.

[10] Lande-Diner, L., Zhang, J., Ben-Porath, I., Amariglio, N., Keshet, I., Hecht, M., Azuara, V., Fisher, A. G., Rechavi, G., & Cedar, H. (2007). Role of DNA methylation in stable gene repression. *Journal of Biological Chemistry*, 282(16), 12194-200.

[11] Cheng, X., & Blumenthal, R. M. (2008). Mammalian DNA methyltransferases: a structural perspective. *Structure*, 16(3), 341-50.

[12] Bostick, M., Kim, J. K., Estève, P. O., Clark, A., Pradhan, S., & Jacobsen, S. E. (2007). UHRF1 plays a role in maintaining DNA methylation in mammalian cells. *Science*, 317(5845), 1760-4.

[13] Sharif, J., Muto, M., Takebayashi, S., Suetake, I., Iwamatsu, A., Endo, T. A., Shinga, J., Mizutani-Koseki, Y., Toyoda, T., Okamura, K., Tajima, S., Mitsuya, K., Okano, M., & Koseki, H. (2007). The SRA protein Np95 mediates epigenetic inheritance by recruiting Dnmt1 to methylated DNA. *Nature*, 450(7171), 908-12.

[14] Goren, A., & Cedar, H. (2003). Replicating by the clock. Nat Rev Mol Cell Biol. *Nature Reviews Molecular Cell Biology*, 4(1), 25-32.

[15] Zhang, J., Xu, F., Hashimshony, T., Keshet, I., & Cedar, H. (2002). Establishment of transcriptional competence in early and late S phase. *Nature*, 420(6912), 198-202.

[16] Dahl, C., Grønbæk, K., & Guldberg, P. (2011). Advances in DNA methylation: 5-hydroxymethylcytosine revisited. *Clinica Chimica Acta*, 412(11-12), 831-6.

[17] Schaefer, M., Pollex, T., Hanna, K., Tuorto, F., Meusburger, M., Helm, M., & Lyko, F. (2010). RNA methylation by Dnmt2 protects transfer RNAs against stress-induced cleavage. Genes Dev. *Genes and Development*, 24(15), 1590-5.

[18] Bestor, TH. (2000). The DNA methyltransferases of mammals. *Human Molecular Genetics*, 9(16), 2395-402.

[19] Jones, P. A., & Liang, G. (2009). Rethinking how DNA methylation patterns are maintained. *Nature Reviews Genetics*, 10(11), 805-11.

[20] Fujita, N., Shimotake, N., Ohki, I., Chiba, T., Saya, H., Shirakawa, M., & Nakao, M. (2000). Mechanism of transcriptional regulation by methyl-CpG binding protein MBD1. *Molecular and Cellular Biology*, 20(14), 5107-18.

[21] Fujita, N., Watanabe, S., Ichimura, T., Tsuruzoe, S., Shinkai, Y., Tachibana, M., Chiba, T., & Nakao, M. (2003). Methyl-CpG binding domain 1 (MBD1) interacts with the Suv39h1-HP1 heterochromatic complex for DNA methylation-based transcriptional repression. *Journal of Biological Chemistry*, 278(26), 24132-8.

[22] Sarraf, S. A., & Stancheva, I. (2004). Methyl-CpG binding protein MBD1 couples histone H3 methylation at lysine 9 by SETDB1 to DNA replication and chromatin assembly. *Molecular Cell*, 15(4), 595-605.

[23] Bogdanović, O., & Veenstra, G. J. (2009). DNA methylation and methyl-CpG binding proteins: developmental requirements and function. *Chromosoma Oct*, 118(5), 549-65, Epub Jun 9.

[24] Hendrich, B., Guy, J., Ramsahoye, B., Wilson, V. A., & Bird, A. (2001). Closely related proteins MBD2 and MBD3 play distinctive but interacting roles in mouse development. *Genes and Development*, 15(6), 710-23.

[25] Hutchins, A. S., Mullen, A. C., Lee, H. W., Sykes, K. J., High, F. A., Hendrich, B. D., Bird, A. P., & Reiner, S. L. (2002). Gene silencing quantitatively controls the function of a developmental trans-activator. *Molecular Cell*, 10(1), 81-91.

[26] Barr, H., Hermann, A., Berger, J., Tsai, H. H., Adie, K., Prokhortchouk, A., Hendrich, B., & Bird, A. (2007). Mbd2 contributes to DNA methylation-directed repression of the Xist gene. *Molecular and Cellular Biology*, 27(10), 3750-7.

[27] Kaji, K., Nichols, J., & Hendrich, B. (2007). Mbd3, a component of the NuRD co-repressor complex, is required for development of pluripotent cells. *Development*, 134(6), 1123-32.

[28] Hendrich, B., Hardeland, U., Ng, H. H., Jiricny, J., & Bird, A. (1999). The thymine gly-
      cosylase MBD4 can bind to the product of deamination at methylated CpG sites. *Na-
      ture,* 401(6750), 301-4.

[29] Millar, C. B., Guy, J., Sansom, O. J., Selfridge, J., MacDougall, E., Hendrich, B.,
      Keightley, P. D., Bishop, S. M., Clarke, A. R., & Bird, A. (2002). Enhanced CpG muta-
      bility and tumorigenesis in MBD4-deficient mice. *Science,* 297(5580), 403-5.

[30] Riccio, A., Aaltonen, L. A., Godwin, A. K., Loukola, A., Percesepe, A., Salovaara, R.,
      Masciullo, V., Genuardi, M., Paravatou-Petsotas, M., Bassi, D. E., Ruggeri- , B. A.,
      Klein-Szanto, A. J., Testa, J. R., Neri, G., & Bellacosa, A. (1999). The DNA repair gene
      MBD4 (MED1) is mutated in human carcinomas with microsatellite instability. *Na-
      ture Genetics,* 23(3), 266-8.

[31] Lewis, J. D., Meehan, R. R., Henzel, W. J., Maurer-Fogy, I., Jeppesen, P., Klein, F., &
      Bird, A. (1992). Purification, sequence, and cellular localization of a novel chromoso-
      mal protein that binds to methylated DNA. *Cell,* 69(6), 905-14.

[32] Adams, V. H., McBryant, S. J., Wade, P. A., Woodcock, C. L., & Hansen, J. C. (2007).
      Intrinsic disorder and autonomous domain function in the multifunctional nuclear
      protein, MeCP2. *Journal of Biological Chemistry,* 282(20), 15057-64.

[33] Chahrour, M., Jung, S. Y., Shaw, C., Zhou, X., Wong, S. T., Qin, J., & Zoghbi, H. Y.
      (2008). MeCP2, a key contributor to neurological disease, activates and represses
      transcription. *Science,* 320(5880), 1224-9.

[34] Young, J. I., Hong, E. P., Castle, J. C., Crespo-Barreto, J., Bowman, A. B., Rose, M. F.,
      Kang, D., Richman, R., Johnson, J. M., Berget, S., & Zoghbi, H. Y. (2005). Regulation
      of RNA splicing by the methylation-dependent transcriptional repressor methyl-CpG
      binding protein 2. *Proceedings of National Academy of Science USA,* 102(49), 17551-8.

[35] Muotri, A. R., Marchetto, M. C., Coufal, N. G., Oefner, R., Yeo, G., Nakashima, K., &
      Gage, F. H. (2010). L1 retrotransposition in neurons is modulated by MeCP2. *Nature,*
      468(7322), 443-6.

[36] Matsui, T., Leung, D., Miyashita, H., Maksakova, I. A., Miyachi, H., Kimura, H., Ta-
      chibana, M., Lorincz, M. C., & Shinkai, Y. (2010). Proviral silencing in embryonic
      stem cells requires the histone methyltransferase ESET. *Nature,* 464(7290), 927-31.

[37] Wyatt, G. R., & Cohen, S. S. (1952). A new pyrimidine base from bacteriophage nu-
      cleic acids. *Nature,* 170(4338), 1072-3.

[38] Penn, N. W., Suwalski, R., O'Riley, C., Bojanowski, K., & Yura, R. (1972). The pres-
      ence of 5-hydroxymethylcytosine in animal deoxyribonucleic acid. *Biochemical Jour-
      nal,* 126(4), 781-90.

[39] Kothari, R. M., & Shankar, V. (1976). Methylcytosine content in the vertebrate deoxy-
      ribonucleic acids: species specificity. *Journal of Molecular Evolution,* 7(4), 325-9.

[40] Kriaucionis, S., & Heintz, N. (2009). The nuclear DNA base 5-hydroxymethylcytosine is present in Purkinje neurons and the brain. *Science*, 324(5929), 929-30.

[41] Tahiliani, M., Koh, K. P., Shen, Y., Pastor, W. A., Bandukwala, H., Brudno, Y., Agarwal, S., Iyer, L. M., Liu, D. R., Aravind, L., & Rao, A. (2009). Conversion of 5-methylcytosine to 5-hydroxymethylcytosine in mammalian DNA by MLL partner TET1. *Science*, 324(5929), 930-5.

[42] Ono, R., Taki, T., Taketani, T., Taniwaki, M., Kobayashi, H., & Hayashi, Y. (2002). LCX, leukemia-associated protein with a CXXC domain, is fused to MLL in acute myeloid leukemia with trilineage dysplasia having t(10;11)(q22;q23). *Cancer Research*, 62(14), 4075-80.

[43] Lorsbach, R. B., Moore, J., Mathew, S., Raimondi, S. C., Mukatira, S. T., & Downing, J. R. (2003). TET1, a member of a novel protein family, is fused to MLL in acute myeloid leukemia containing the t(10;11)(q22;q23). *Leukemia*, 17(3), 637-41.

[44] Ito, S., D'Alessio, A. C., Taranova, O. V., Hong, K., Sowers, L. C., & Zhang, Y. (2010). Role of Tet proteins in 5mC to 5hmC conversion, ES-cell self-renewal and inner cell mass specification. *Nature*, 466(7310), 1129-33.

[45] Inoue, A., & Zhang, Y. (2011). Replication-dependent loss of 5-hydroxymethylcytosine in mouse preimplantation embryos. *Science*, 334(6053), 194.

[46] Song, C. X., Szulwach, K. E., Fu, Y., Dai, Q., Yi, C., Li, X., Li, Y., Chen, C. H., Zhang, W., Jian, X., Wang, J., Zhang, L., Looney, T. J., Zhang, B., Godley, L. A., Hicks, L. M., Lahn, B. T., Jin, P., & He, C. (2011). Selective chemical labeling reveals the genomewide distribution of 5-hydroxymethylcytosine. *Nature Biotechnology*, 29(1), 68-72.

[47] Abdel-Wahab, O., Mullally, A., Hedvat, C., Garcia-Manero, G., Patel, J., Wadleigh, M., Malinge, S., Yao, J., Kilpivaara, O., Bhat, R., Huberman, K., Thomas, S., Dolgalev, I., Heguy, A., Paietta, E., Le Beau, M. M., Beran, M., Tallman, M. S., Ebert, B. L., Kantarjian, H. M., Stone, R. M., Gilliland, D. G., Crispino, J. D., & Levine, R. L. (2009). Genetic characterization of TET1, TET2, and TET3 alterations in myeloid malignancies. *Blood*, 114(1), 144-7.

[48] Ko, M., Huang, Y., Jankowska, A. M., Pape, U. J., Tahiliani, M., Bandukwala, H. S., An, J., Lamperti, E. D., Koh, K. P., Ganetzky, R., Liu, X. S., Aravind, L., Agarwal, S., Maciejewski, J. P., & Rao, A. (2010). Impaired hydroxylation of 5-methylcytosine in myeloid cancers with mutant TET2. *Nature*, 468(7325), 839-43.

[49] Krais, A. M., Park, Y. J., Plass, C., & Schmeiser, H. H. (2011). Determination of genomic 5-hydroxymethyl-2'-deoxycytidine in human DNA by capillary electrophoresis with laser induced fluorescence. *Epigenetics*, 6(5), 560-5.

[50] Wu, H., & Zhang, Y. (2011). Mechanisms and functions of Tet protein-mediated 5-methylcytosine oxidation. *Genes and Development*, 25(23), 2436-52.

[51] Lande-Diner, L., Zhang, J., & Cedar, H. (2009). Shifts in replication timing actively affect histone acetylation during nucleosome reassembly. *Molecular Cell*, 34(6), 767-74.

[52] Hashimoto, H., Vertino, P. M., & Cheng, X. (2010). Molecular coupling of DNA methylation and histone methylation. *Epigenomics*, 2(5), 657-69.

[53] Goren, A., Tabib, A., Hecht, M., & Cedar, H. (2008). DNA replication timing of the human beta-globin domain is controlled by histone modification at the origin. *Genes and Development*, 22(10), 1319-24.

[54] Laurent, L., Wong, E., Li, G., Huynh, T., Tsirigos, A., Ong, C. T., Low, H. M., Kin, Sung. K. W., Rigoutsos, I., Loring, J., & Wei, C. L. (2010). Dynamic changes in the human methylome during differentiation. Genome Res. *Genome Research*, 20(3), 320-31.

[55] Fuks, F., Hurd, P. J., Deplus, R., & Kouzarides, T. (2003). The DNA methyltransferases associate with HP1 and the SUV39H1 histone methyltransferase. *Nucleic Acids Research*, 31(9), 2305-12.

[56] Ohki, I., Shimotake, N., Fujita, N., Jee, J., Ikegami, T., Nakao, M., & Shirakawa, M. (2001). Solution structure of the methyl-CpG binding domain of human MBD1 in complex with methylated DNA. *Cell*, 105(4), 487-97.

[57] Ho, K. L., McNae, I. W., Schmiedeberg, L., Klose, R. J., Bird, A. P., & Walkinshaw, M. D. (2008). MeCP2 binding to DNA depends upon hydration at methyl-CpG. *Molecular Cell*, 29(4), 525-31.

[58] Grewal, S. I., & Jia, S. (2007). Heterochromatin revisited. *Nature Reviews Genetics*, 8(1), 35-46.

[59] Tachibana, M., Matsumura, Y., Fukuda, M., Kimura, H., & Shinkai, Y. (2008). G9a/GLP complexes independently mediate H3K9 and DNA methylation to silence transcription. *EMBO Journal*, 27(20), 2681-90.

[60] Milne, T. A., Briggs, S. D., Brock, H. W., Martin, M. E., Gibbs, D., Allis, C. D., & Hess, J. L. (2002). MLL targets SET domain methyltransferase activity to Hox gene promoters. *Molecular Cell*, 10(5), 1107-17.

[61] Birke, M., Schreiner, S., García-Cuéllar, M. P., Mahr, K., Titgemeyer, F., & Slany, R. K. (2002). The MT domain of the proto-oncoprotein MLL binds to CpG-containing DNA and discriminates against methylation. *Nucleic Acids Research*, 30(4), 958-65.

[62] Xin, Z., Tachibana, M., Guggiari, M., Heard, E., Shinkai, Y., & Wagstaff, J. (2003). Role of histone methyltransferase G9a in CpG methylation of the Prader-Willi syndrome imprinting center. *Journal of Biological Chemistry*, 278(17), 14996-5000.

[63] Fan, S., Zhang, M. Q., & Zhang, X. (2008). Histone methylation marks play important roles in predicting the methylation status of CpG islands. *Biochemistry Biophysical Research Communications*, 374(3), 559-64.

[64] Wang, J., Hevi, S., Kurash, J. K., Lei, H., Gay, F., Bajko, J., Su, H., Sun, W., Chang, H., Xu, G., Gaudet, F., Li, E., & Chen, T. (2009). The lysine demethylase LSD1 (KDM1) is required for maintenance of global DNA methylation. Nat Genet. *Nature Genetics*, 41(1), 125-9.

[65] Jones, P. L., Veenstra, G. J., Wade, P. A., Vermaak, D., Kass, S. U., Landsberger, N., Strouboulis, J., & Wolffe, A. P. (1998). Methylated DNA and MeCP2 recruit histone deacetylase to repress transcription. *Nat Genet. Nature Genetics*, 19(2), 187-91.

[66] Nan, X., Ng, H. H., Johnson, C. A., Laherty, C. D., Turner, B. M., Eisenman, R. N., & Bird, A. (1998). Transcriptional repression by the methyl-CpG-binding protein MeCP2 involves a histone deacetylase complex. *Nature*, 393(6683), 386-9.

[67] Ooi, S. K., Qiu, C., Bernstein, E., Li, K., Jia, D., Yang, Z., Erdjument-Bromage, H., Tempst, P., Lin, S. P., Allis, C. D., Cheng, X., & Bestor, T. H. (2007). DNMT3L connects unmethylated lysine 4 of histone H3 to de novo methylation of DNA. *Nature*, 448(7154), 714-7.

[68] Albrecht, U., Sutcliffe, J. S., Cattanach, B. M., Beechey, C. V., Armstrong, D., Eichele, G., & Beaudet, A. L. (1997). Imprinted expression of the murine Angelman syndrome gene, Ube3a, in hippocampal and Purkinje neurons. *Nature Genetics*, 17(1), 75-8.

[69] Kishino, T., Lalande, M., & Wagstaff, J. (1997). UBE3A/E6-AP mutations cause Angelman syndrome. *Nature Genetics*, 15(1), 70-3.

[70] Xue, F., Tian, X. C., Du, F., Kubota, C., Taneja, M., Dinnyes, A., Dai, Y., Levine, H., Pereira, L. V., & Yang, X. (2002). Aberrant patterns of X chromosome inactivation in bovine clones. *Nature Genetics*, 31(2), 216-20.

[71] Okano, M., Bell, D. W., Haber, D. A., & Li, E. (1999). DNA methyltransferases Dnmt3a and Dnmt3b are essential for de novo methylation and mammalian development. *Cell*, 99(3), 247-57.

[72] Shirohzu, H., Kubota, T., Kumazawa, A., Sado, T., Chijiwa, T., Inagaki, K., Suetake, I., Tajima, S., Wakui, K., Miki, Y., Hayashi, M., Fukushima, Y., & Sasaki, H. (2002). Three novel DNMT3B mutations in Japanese patients with ICF syndrome. *American Journal of Medical Genetics*, 112(1), 31-7.

[73] Kubota, T., Furuumi, H., Kamoda, T., Iwasaki, N., Tobita, N., Fujiwara, N., Goto, Y., Matsui, A., Sasaki, H., & Kajii, T. (2004). ICF syndrome in a girl with DNA hypomethylation but without detectable DNMT3B mutation. *American Journal of Medical Genetics*, 129A(3), 290-3.

[74] De Marzo, A. M., Marchi, V. L., Yang, E. S., Veeraswamy, R., Lin, X., & Nelson, W. G. (1999). Abnormal regulation of DNA methyltransferase expression during colorectal carcinogenesis. *Cancer Research*, 59(16), 3855-60.

[75] Roll, J. D., Rivenbark, A. G., Jones, W. D., & Coleman, W. B. (2008). DNMT3b overexpression contributes to a hypermethylator phenotype in human breast cancer cell lines. *Molecular Cancer*, 7, 15.

[76] Saito, Y., Kanai, Y., Nakagawa, T., Sakamoto, M., Saito, H., Ishii, H., & Hirohashi, S. (2003). Increased protein expression of DNA methyltransferase (DNMT) 1 is significantly correlated with the malignant potential and poor prognosis of human hepatocellular carcinomas. *International Journal of Cancer*, 105(4), 527-32.

[77] Amir, R. E., Van den Veyver, I. B., Wan, M., Tran, C. Q., Francke, U., & Zoghbi, H. Y. (1999). Rett syndrome is caused by mutations in X-linked MECP2, encoding methyl-CpG-binding protein 2. *Nature Genetics,* 23(2), 185-8.

[78] Chunshu, Y., Endoh, K., Soutome, M., Kawamura, R., & Kubota, T. (2006). A patient with classic Rett syndrome with a novel mutation in MECP2 exon 1. *Clinical Genetics,* 70(6), 530-1.

[79] Martinowich, K., Hattori, D., Wu, H., Fouse, S., He, F., Hu, Y., Fan, G., & Sun, Y. E. (2003). DNA methylation-related chromatin remodeling in activity-dependent BDNF gene regulation. *Science,* 302(5646), 890-3.

[80] Miyake, K., Hirasawa, T., Soutome, M., Itoh, M., Goto, Y., Endoh, K., Takahashi, K., Kudo, S., Nakagawa, T., Yokoi, S., Taira, T., Inazawa, J., & Kubota, T. (2011). The protocadherins, PCDHB1 and PCDH7, are regulated by MeCP2 in neuronal cells and brain tissues: implication for pathogenesis of Rett syndrome. *BMC Neuroscience,* 12, 81.

[81] Clayton-Smith, J., O'Sullivan, J., Daly, S., Bhaskar, S., Day, R., Anderson, B., Voss, A. K., Thomas, T., Biesecker, L. G., Smith, P., Fryer, A., Chandler, K. E., Kerr, B., Tassabehji, M., Lynch, S. A., Krajewska-Walasek, M., Mc Kee, S., Smith, J., Sweeney, E., Mansour, S., Mohammed, S., Donnai, D., & Black, G. (2011). Whole-exome-sequencing identifies mutations in histone acetyltransferase gene KAT6B in individuals with the Say-Barber-Biesecker variant of Ohdo syndrome. *American Journal of Human Genetics,* 89(5), 675-81.

[82] Kleefstra, T., Brunner, H. G., Amiel, J., Oudakker, A. R., Nillesen, W. M., Magee, A., Geneviève, D., Cormier-Daire, V., van Esch, H., Fryns, J. P., Hamel, B. C., Sistermans, E. A., de Vries, B. B., & van Bokhoven, H. (2006). Loss-of-function mutations in euchromatin histone methyl transferase 1 (EHMT1) cause the 9q34 subtelomeric deletion syndrome. *American Journal of Human Genetics,* 79(2), 370-7.

[83] Weaver, I. C., Cervoni, N., Champagne, F. A., D'Alessio, A. C., Sharma, S., Seckl, J. R., Dymov, S., Szyf, M., & Meaney, MJ. (2004). Epigenetic programming by maternal behavior. *Nature Neuroscience,* 7(8), 847-54.

[84] Lillycrop, K. A., Phillips, E. S., Jackson, A. A., Hanson, M. A., & Burdge, G. C. (2005). Dietary protein restriction of pregnant rats induces and folic acid supplementation prevents epigenetic modification of hepatic gene expression in the offspring. *Journal of Nutrition,* 135(6), 1382-6.

[85] Lillycrop, K. A., Phillips, E. S., Torrens, C., Hanson, M. A., Jackson, A. A., & Burdge, G. C. (2008). Feeding pregnant rats a protein-restricted diet persistently alters the methylation of specific cytosines in the hepatic PPAR alpha promoter of the offspring. *British Journal of Nutrition,* 100(2), 278-82.

[86] Heijmans, B. T., Tobi, E. W., Stein, A. D., Putter, H., Blauw, G. J., Susser, E. S., Slagboom, P. E., & Lumey, L. H. (2008). Persistent epigenetic differences associated with prenatal exposure to famine in humans. *Proceedings of National Academy of Science USA,* 105(44), 17046-9.

[87] Tobi, E. W., Lumey, L. H., Talens, R. P., Kremer, D., Putter, H., Stein, A. D., Slag-boom, P. E., & Heijmans, B. T. (2009). DNA methylation differences after exposure to prenatal famine are common and timing- and sex-specific. *Human Molecular Genetics*, 18(21), 4046-53.

[88] Franklin, T. B., Russig, H., Weiss, I. C., Gräff, J., Linder, N., Michalon, A., Vizi, S., & Mansuy, I. M. (2010). Epigenetic transmission of the impact of early stress across generations. *Biological Psychiatry*, 68(5), 408-15.

[89] Horsthemke, B. (2007). Heritable germline epimutations in humans. *Nature Genetics*, 39(5), 573-4.

[90] Daxinger, L., & Whitelaw, E. (2010). Transgenerational epigenetic inheritance: more questions than answers. *Genome Research*, 20(12), 1623-8.

[91] Seong, K. H., Li, D., Shimizu, H., Nakamura, R., & Ishii, S. (2011). Inheritance of stress-induced, ATF-2-dependent epigenetic change. *Cell*, 145(7), 1049-61.

[92] De Gobbi, M., Anguita, E., Hughes, J., Sloane-Stanley, J. A., Sharpe, J. A., Koch, C. M., Dunham, I., Gibbons, R. J., Wood, W. G., & Higgs, D. R. (2007). Tissue-specific histone modification and transcription factor binding in alpha globin gene expression. *Blood*, 110(13), 4503-10.

[93] Yagi, S., Hirabayashi, K., Sato, S., Li, W., Takahashi, Y., Hirakawa, T., Wu, G., Hattori, N., Hattori, N., Ohgane, J., Tanaka, S., Liu, X. S., & Shiota, K. (2008). DNA methylation profile of tissue-dependent and differentially methylated regions (T-DMRs) in mouse promoter regions demonstrating tissue-specific gene expression. *Genome Research Dec*, 18(12), 1969-78.

[94] Beck, S. (2010). Taking the measure of the methylome. *Nature Biotechnology*, 28(10), 1026-8.

[95] Kubota, T., Miyake, K., & Hirasawa, T. (2011). Epigenetic modifications: genetic basis of environmental stress response. *In DNA Replication / Book 1. Rijeka.*, Intech, http://www.intechopen.com/books/fundamental-aspects-of-dna-replication/epigenetic-modifications-genetic-basis-of-environmental-stress-response, accessed 15 July 2012.

Chapter 4

Roles of Methylation and Sequestration in the Mechanisms of DNA Replication in some Members of the Enterobacteriaceae Family

Amine Aloui, Alya El May,
Saloua Kouass Sahbani and Ahmed Landoulsi

Additional information is available at the end of the chapter

1. Introduction

When growing cells divide, they need to copy their genetic material and distribute it to en-sure that each daughter cell receives one copy. This is a challenging task especially when the enormous length of the DNA compared to the cell size is considered. During DNA replica-tion, organization of the chromosomes is even more demanding, since replication forks con-tinuously produce new DNA. This DNA contains all the information required to build the cells and tissues of a prokaryotic or an eukaryotic organism. The exact replication of this in-formation in any species assures its genetic continuity from generation to generation and is critical to the normal development of an individual. The information stored in DNA is ar-ranged in hereditary units known as genes that control the identifiable traits of an organism. Discovery of the structure of DNA and subsequent elucidation of how DNA directs synthe-sis of RNA, which then directs assembly of proteins -the so-called central dogma - were monumental achievements that marked the early days of molecular biology. However, the simplified representation of the central dogma as DNA → RNA → protein does not reflect the role of proteins in the synthesis of nucleic acids. Moreover, proteins are largely responsi-ble for regulating DNA replication and gene expression, the entire process whereby the in-formation encoded in DNA is decoded into the proteins that characterize various cell types. Two of these proteins are the DNA adenine methyltransferase (Dam) and the DNA-Binding Protein (SeqA).

1.1. The Dam methyltransferase

Methylation of DNA by the Dam methyltransferase provides an epigenetic signal that influences and regulates numerous physiological processes in the bacterial cell, including chromosome replication, mismatch repair, transposition, and transcription. A growing number of reports ascribed a role to DNA adenine methylation in regulating the mechanisms of DNA replication in diverse pathogens like in *Salmonella* Typhimurium and *Escherichia coli*, ...), suggesting that DNA methylation may be a widespread and versatile regulator of this process. The Dam enzyme catalyzes the transfer of a methyl group from S -adenosyl-L-methionine (SAM) to the N^6 position of the adenine residue in GATC sequences, using base flipping to position the base in the enzyme's catalytic site (Figure 1).

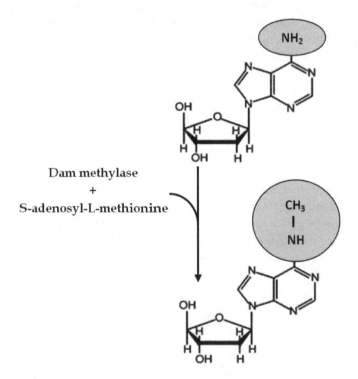

Figure 1. Dam catalyzes the transfer of a methyl group from S-adenosyl-L-methionine to the N^6 position of adenine.

The natural substrate for the enzyme is hemimethylated DNA, where one strand is methylated and the other is not. This is the configuration of DNA immediately behind the replication fork. Double stranded DNA is a better methyl receiver than denatured DNA, and there is little difference in the rate of methylation between unmethylated and hemimethylated DNA [28]. The Dam enzyme appears to have two SAM binding sites; one is the catalytic site and the other increases specific binding to DNA, probably through an allosteric transformation [13]. Dam is thought to bind the template and to slide processing along the DNA, methylating about 55 GATC sites per binding event [60]. There are about 130 molecules of Dam per *Escherichia coli* cell, and this is considered optimal because it allows a period of time between the synthesis of the extending nucleotide chains and the methylation of the GATC sequences within them [15]. The cellular level of Dam is regulated mainly by transcription; any increases or decreases in the number of Dam molecules can profoundly alter the physiological properties of the cell.

Dam competes with two other proteins, MutH and SeqA, for hemimethylated GATC substrate sites. These two proteins act before Dam to participate in the removal of replication errors (MutH) and to form the compact and properly super coiled chromosome structure for the nucleoid (SeqA). Increasing the cellular level of Dam causes a decrease in the amount of hemimethylated DNA, and prevents these two proteins from carrying out their functions, leading to an increased mutation rate and a change in the super coiling of the chromosome, respectively [27; 42; 40]. Although Dam methylase is a highly processing enzyme, it may become less processing at GATC sites, flanked by specific DNA sequences [50]. Reduced rate of processing may allow for a competition between Dam and specific DNA- binding proteins, thus permitting the formation of non-methylated GATCs which depended on the growth phase and the growth rate, suggesting that the proteins that bind to them could be involved in gene expression or in the maintenance of chromosome structure. The unmethylated dam sites appear to be mostly [53] or completely [49] modified in strains overproducing Dam, suggesting that the enzyme competes with other DNA - binding proteins at these specific sites. In addition to the unmethylated GATC sites discussed above, persistent hemimethylated sequences have been detected in the chromosome [48; 18]. These are distinct from the transiently hemimethylated GATC sites that occur immediately behind the replication fork due to the time lag between DNA replication and Dam methylation.

1.2. The SeqA protein

SeqA protein was discovered in some prokaryotes as a protein involved in the methylation / hemimethylation cycle of DNA replication [41]. This protein regulates the activation of the chromosome replication origin [41]. Experiments have shown that SeqA has a high affinity to hemimethylated as compared to fully methylated DNA. It binds specifically to GATC sequences which are methylated on the Adenine of the old strand but not on the new strand. Such hemimethylated DNA is produced by progression of the replication forks and lasts until Dam methyltransferase methylates the new strand (Figure 2). It is therefore believed that a region of hemimethylated DNA covered by SeqA follows the replication fork.

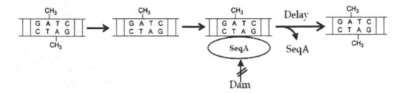

Figure 2. Helically phased GATC sites can be bound by SeqA when they are in the hemimethylated state. Binding of SeqA inhibits Dam methylation, maintaining the hemimethylated state for a portion of the cell cycle. Dissociation of SeqA allows Dam to methylate the hemimethylated DNA, thus generating fully methylated DNA.

Proper chromosome segregation also requires SeqA [5]. Furthermore, SeqA trails the DNA replication fork and may contribute to nucleic organization in newly replicated DNA [17; 37; 40; 67]. Aside from its roles in chromosome replication and nucleic segregation, SeqA is known to regulate the transcription of certain genes. In bacteriophage lambda, SeqA activates the p_R promoter in a GATC methylation-dependent fashion. SeqA also acts as a transcriptional co-activator by facilitating binding of the cII transcription factor to the lambda p_I and p_{aQ} promoters. Competition between SeqA and the OxyR repressor for hemimethylated GATC sites has been shown to regulate phase variation in the *Escherichia coli agn43* gene [51]. These examples raised the possibility that SeqA binding to critical GATC sites may likewise regulate the expression of prokaryotic genes like in *Escherichia coli* and *Salmonella*, which are members of the *Enterobacteriaceae* family.

1.3. The competition of Dam methylase and SeqA for GATC sites

Newly synthesized hemimethylated DNA is also a target for Dam methylase. In fact, Dam and SeqA have been suggested to be in competition for hemimethylated DNA [34]. Experiments with unsynchronized cells indicate that the sequestration period becomes shorter upon Dam over-expression [15; 57]. SeqA binding is largely limited to hemimethylated DNA, and the action of Dam will, therefore, transform DNA into a non-target for SeqA. This protein was able to bind DNA despite Dam overproduction. Recent findings showed that SeqA bound to DNA was not actively dissociated by Dam methylase [34]. The same study showed that the SeqA protein spontaneously dissociated from bound DNA after some minutes *in vitro* and that re-binding to the same site was inhibited by methylation [34]. We reasoned that *in vivo* system, the effect of Dam overproduction on SeqA re-binding should increase with increasing distance from replication forks towards the origin of chromosomal replication. This is because the longer a SeqA molecule was bound to the DNA the more likely is its dissociation. Such an effect might be too small to be observed by visual comparison of SeqA binding patterns. This shows that Dam and SeqA are in continuous competition for GATC sites *in vivo* with SeqA being the considerably stronger competitor. Since SeqA has been shown to bind better to DNA regions with more densely packed GATC sites, we speculated that such regions would allow SeqA to be better in competing against Dam than do regions with fewer GATC sites. Recent data indicate that differences in GATC density have only minor impact on the competition of SeqA against Dam methyltransferase [32].

2. DNA replication

The initiation of chromosomal replication occurs only once during the cell cycle in both pro-karyotes and eukaryotes. This initiation is the first and tightly controlled step of a DNA syn-thesis. Because much of what is known about the regulation of the initiation of bacterial chromosomal replication comes from studies of *Escherichia coli* and *Salmonella enterica* sero-var Typhimurium, this review focuses mainly on regulatory mechanisms in these species.

In prokaryotic cells, DNA replication and segregation are not temporally separated process-es. Some evidence suggests that newly synthesized DNA is continuously segregated to op-posite cellular positions [45; 52]. Other work indicates that some parts of segregation may be more abrupt and domain specific [7; 21]. Coordination of DNA replication and chromosome segregation is complicated by the ability of growing with overlapping replication cycles [19; 55]. Whereas during slow growth, chromosomes are replicated in a simple pattern with only one pair of forks; replication during fast growth occurs with one pair of old and two pairs of new forks on one chromosome. (Forks are considered to be 'old forks' as soon as new forks appear at initiation.) Depending on the exact conditions, a cell can have four copies of the multi-fork chromosome and a total of 24 replication forks per cell [22; 44]. How the cell meets the obvious need for efficient organization during such extensive replication is largely unknown. However, the SeqA protein is one of the strongest candidates to contribute [41; 61]. Loss of SeqA leads to severe growth impairment during the rapid but not slow growth [16]. Biochemical studies established that SeqA binding is specific to the sequence GATC with high preference for hemimethylated over fully methylated DNA [10; 58; 11; 34]. Hemi-methylation occurs at newly replicated GATC sites which have not yet been re-methylated by the Dam methylase. A transient hemimethylation after the passage of the replication fork was found in an analysis of 10 individual GATC sites [18]. Similarly, transient binding of SeqA was detected at seven genomic sites with multiple GATC sequences [67]. Multiple DNA- bound SeqA dimers can oligomerize to form a higher order structure [25; 47]. The above findings suggested that a SeqA complex follows the replication forks, potentially in a tread milling fashion, growing at the leading end and diminishing at the tailing end. The re-duction of the SeqA bound region at the most replisome-distant GATCs would come about through the activity of Dam which turns these sites into non-targets for SeqA by its methyla-tion activity. The process described above is called DNA sequestration.

3. DNA methylation by Dam protein

As mentioned above, the Dam methyltransferase of *Enterobacteriaceae* methylates adenine at the N^6 position in GATC sequences. Methylation of DNA has multiple consequences con-cerning bacterial physiology including the regulation of chromosome replication, DNA seg-regation, mismatch repair, transposition, and transcriptional regulation. The molecular basis for the pleiotropic phenotypes associated with Dam is the differential methylation of DNA resulting in an altered affinity of regulatory DNA-binding proteins. Regulatory proteins might preferentially bind to non -methylated DNA, thereby blocking methylation by Dam, while other proteins bind with high affinity only to hemimethylated or fully methylated DNA [65]. Therefore, it is not surprising that Dam has an impact on pathogenesis, virulence gene expression, influences DNA replication [2] and many other processes [4] in *Salmonella* microorganisms. Dam-overproducing (Dam^{OP}) as well as *dam* mutant strains have been used to assess the role of DNA methylation in DNA replication. By using these strains it is possible to alter the methylation pattern in regulatory regions of genes, thereby changing the binding affinity for regulatory proteins. Although Dam^{OP} does not reflect a physiologically relevant condition, as the Dam levels have been found to remain basically constant in the cell, it is a functional tool to analyze the effects of changes in DNA methylation patterns on gene expression. Strategies of this kind have been successfully used for decades in both eukaryotes and prokaryotes to perturb gene regulation for experimental purposes.

4. DNA sequestration by SeqA protein

Replication of the bacterial chromosomal DNA initiates only once, at a specific region known as the origin of chromosomal replication (*oriC*), by the initiator protein DnaA. This protein interacts specifically with 9-bp non-palindromic sequences (DnaA boxes) that exists at *oriC*. To ensure that initiation at an origin occurs only once per cell cycle, specific mechanisms exist to control chromosomal replication. In one mechanism, the SeqA protein that is tightly bound to hemimethylated DNA by a mechanism known as sequestration and which recognizes GATC sequences overrepresented within *oriC* and prefers binding to hemimethylated over fully or unmethylated *oriC*.

The chromosomal DNA is methylated at adenine residues in GATC sequences by Dam methylase. Following passage of the DNA replication fork, GATC sites methylated on the top and bottom strands in a mother cell (denoted as fully methylated) are converted into two hemimethylated DNA duplexes: one methylated on the top strand and non-methylated on the bottom strand and one methylated on the bottom strand and non-methylated on the top strand due to semi-conservative replication. Most GATC sites are rapidly re-methylated by the enzyme Dam methylase and exist in the hemimethylated state for only a fraction of the cell cycle (Figure 3).

Exceptions to this are the DNA replication origin of *Escherichia coli* and *Salmonella typhimurium*, the dnaA promoter, and possibly additional GATC sites in the chromosome which bind SeqA. SeqA preferentially binds to clusters of two or more hemimethylated GATC sites spaced one to two helical turns apart (Figure 4).

In the case of *oriC*, sequestration delays re-methylation and prevents binding of the DnaA protein, which controls the initiation of DNA replication. At other sites, binding of SeqA tetramers to hemimethylated GATC sites may organize nucleic domains. Notably, the transcription profile of a SeqA mutant was found to be similar to that of a Dam overproducer strain. Based on this observation, a model was developed in which Dam and SeqA compete for binding to hemimethylated for binding to hemimethylated DNA generated at the replication fork.

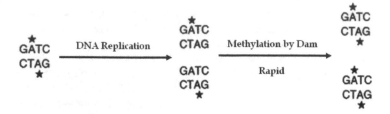

Figure 3. The vast majority of chromosomal GATC sites are fully methylated until DNA replication generates two hemimethylated species, one methylated on the top strand and one methylated on the bottom strand. Within a short time after replication (less than 5 minutes), Dam methylates the nonmethylated GATC site, regenerating a fully methylated GATC site.

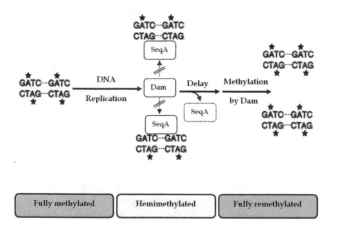

Figure 4. Two or more helically phased GATC sites can be bound by SeqA when they are in the hemimethylated state. Binding of SeqA inhibits Dam methylation, maintaining the hemimethylated state for a portion of the cell cycle. Dissociation of SeqA allows Dam to methylate the hemimethylated DNAs, generating fully methylated DNA.

4.1. Binding of SeqA to Hemimethylated GATC Sequences

As mentioned above, the adenine residues of GATC sequences are methylated on their 6 amino group by Dam methyltransferase [23; 18]. Upon replication, the GATC sequences on the newly replicated strand remain transiently unmethylated, leading to a hemimethylated state of the DNA duplex. The new strand is subsequently methylated by Dam, and the duplex becomes methylated on both strands. The initiation of chromosome replication at the origin of chromosomal replication *oriC*, which contains repeated GATC sequences, is tightly controlled [43; 38]. Once initiation is fired, reinitiation from the newly formed *oriC* is prevented by an *oriC* sequestration process affected by the binding of SeqA protein to the newly replicated, hemi-methylated origin [41; 34]. The hemimethylated state of the replicated *oriC* is maintained for about one-third of the cell cycle, whereas it persists in other chromosomal regions for at most 2 minutes. Further, the asynchronous and overinitiation of chromosomal replication characteristic of *seqA* mutants indicates that SeqA is a negative modulator of chromosomal initiation at *oriC* [41, 16].

Topoisomerase IV is essential for the de-catenation and segregation of replicated chromosomes at cell division [35; 36; 1; 68; 29]. Together with DNA gyrase, it also removes the positive super coils that accumulate in front of replication forks and growing mRNA transcripts. SeqA has been shown to promote the relaxation and the de-catenation activity of topoisomerase IV [33]. This appears to result from a specific interaction between topoisomerase IV and SeqA. Besides the asynchrony and overinitiation of chromosomal replication, *seqA* mutants have an aberrant nucleic structure, an increased frequency of abnormal segregation, and increased negative superhelicity of chromosomal and plasmid DNA [41; 30; 63; 64]. These findings suggest that interaction with SeqA is required for proper functioning of topoisomerase IV *in vivo*. In addition, SeqA functions as a transcriptional regulator of the bacteriophage *pR* promoter [59].

The C-terminal region of SeqA interacts via hydrogen bonds and van der Waals contacts with the major groove of DNA, with the hemimethylated A-T base pair and also with the surrounding bases and DNA backbone. The Nuclear Magnetic Resonance (NMR) structure of hemimethylated GATC revealed that it has an unusual backbone structure and a remarkably narrow major groove and suggested that this peculiar structural feature might contribute to recognition of hemimethylated GATC sites by SeqA protein [6]. To form a stable SeqA-DNA complex in the presence of competitor DNA, one SeqA tetramer binds to each of two hemimethylated GATC sequences [26] that are up to 31 bases apart on the DNA [17]. The sequential binding of SeqA tetramers to hemimethylated sites leads to the formation of higher order complexes [26]. Further, the binding of SeqA proteins to at least six adjacent hemimethylated sites induces the aggregation of free proteins onto the bound proteins, thus implying cooperative interaction between the SeqA proteins.

4.2. Effects of *seqA* disruption on DNA replication

As we said before, following the replication fork progression and the nascent strand synthesis, the daughter DNA becomes hemimethylated. SeqA protein binds to the hemimethylated GATC sequences (hemi-sites) and performs various roles to control the cell cycle progression. Immediately after the initiation of replication SeqA binds to the replicated *oriC* and sequesters it from remethylation and re-initiation of replication at the replicated *oriC*. SeqA tracks replication forks as a multiprotein complex and contributes to the maintenance of superhelicity and de-catenation of daughter chromosomes through the stimulation of topoiso-merase IV and results in a synchronous replication.

When rounds of replication are allowed to run to completion, the number of chromosomes per cell is 2n (n = 0, 1, 2, 3, etc). When initiations are asynchronous, as in *dnaA* (Ts) initiation mutants at the permissive temperature and in the *Escherichia coli dam* mutant [14; 56], the presence of a different number of chromosome equivalents (three, five, six, etc.) was detect-ed by flow cytometry. The presence of cells containing a number of chromosomes different from 2n suggests that the *seqA* mutant has a defect in the synchrony of replication initiation. Wild type and *seqA* mutant of *Salmonella enterica* serovar Typhimurium growing exponen-tially in glucose–casamino acid medium were treated with rifampicin and cephalexin, which block initiation of replication and cell division respectively. Wild-type cells initiated replica-tion synchronously (number of chromosomes per cell is 2n). The appearance of cells with chromosome numbers other than 2n indicates a moderate asynchrony of initiation.

So, the flow cytometer analysis of a *seqA* mutants has shown that replication initiation is asynchronous and can occur throughout the cell cycle, not only at the normal cell age for initiation. The most likely reason for this asynchrony phenotype is that secondary initiations occurred at newly replicated origins in *seqA* mutants, due to lack of sequestration and inade-quate methylation. We showed that the initiation synchrony was dependent on intact GATC methylation sites.

This loss of synchrony affected culture growth rates and cell size distributions only slightly and suggest that *seqA* mutants have a slight defect in synchronizing replication initiation. All these results suggest that DNA sequestration plays a role in preventing the occurrence of multiple initiations at a single origin in the same replication cycle. However, using flow cy-tometry, we found that the asynchrony of initiation, which is one of the phenotypes of the *seqA* mutation, was returned to almost normal in a *seqA* null mutant harbouring the wild-type *seqA* gene under the control of a *tac-* promoter.

The OFF to ON phase rate was reduced in a *seqA* mutant, but much of this effect could be accounted for by a reduction in the Dam / DNA ratio caused by increased asynchronous ini-tiation of DNA replication that occurs in the absence of SeqA, which normally sequesters *oriC* and plays a critical role in timing of DNA replication [8].

5. Membrane sequestration hemimethylated of *oriC*

The coordination of the synchronization of the replication initiation, the activation of the DnaA protein at *oriC*, and the cellular cycle suggested the existence of a very narrow interac-tion between the bacterial membrane and the SeqA protein [39]. Early studies demonstrated that membranes are capable of binding to hemimethylated *oriC in vitro* and *in vivo*, but not to fully methylated or unmethylated *oriC* [48]. While they are sequestered at the membrane, the recently replicated origins are unavailable for re-initiation and are protected from methylation by Dam methylase for an extended period. The origins remain sequestered until conditions in the cell are no longer in a state supportive for initiation (Figure 5).

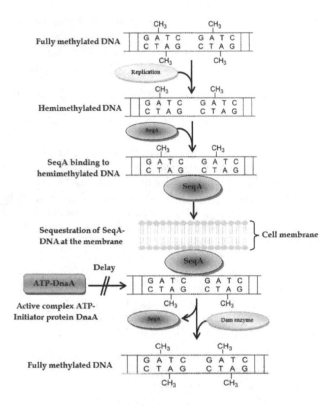

Figure 5. Membrane sequestration of recently replicated origins [4].

Prior to initiation of DNA replication, Dam methylase sites are fully methylated. Immediately following replication, the newly synthesized strand is unmethylated, and the resulting hemimethylated origin is sequestered at the lipid bilayer of membrane by SeqA. This is not accessible to replicatively active ATP–DnaA. After approximately one-third of the cell cycle, the sequestered origin is released and methylated by Dam methylase. At this point in the cell cycle, the levels of ATP–DnaA are not sufficient to catalyze a new round of replication. As such, sequestration serves as a mechanism to prevent secondary initiations. Subsequent work identified SeqA protein to be an essential factor in the *oriC* sequestration. Even though the first steps of SeqA purification involve liberating SeqA from the membrane fraction of cell lysates by treatment with high concentrations of salt and sonication, the primary sequence for SeqA protein does not suggest any obvious membrane-associating domains. This is supported by the crystal structure of the C-terminal DNA-binding domain, and by biochemical studies that show that the N-terminal domain serves in the aggregation of SeqA protein into functional homotetramers [24]. Yet, there is some evidence that SeqA has an association with membranes [20; 62]. The original data that newly replicated, hemimethylated origins are sequestered at the membrane hold true. Whether the membrane sequestration of *oriC* occurs directly through the SeqA protein or through an as yet unidentified factor remains unclear.

6. Cooperation between "Dam" and "SeqA" in DNA Replication

In *Escherichia coli*, persistent hemimethylated sites have been detected at the origin of chromosome replication, *oriC*, and the region surrounding it [18]. This region includes the dnaA gene, which is located 43 kb from *oriC*. DnaA initiates chromosome replication by binding to *oriC* and facilitating duplex opening to load DnaB helicase and DNA polymerase III holoenzyme. The persistence of the hemimethylated state is due to the high density of GATC sequences in *oriC* (11 in 245 bp) and in the promoter region of *dnaA* (8 in 219 bp), providing multiple binding sites for the SeqA protein. The SeqA induced hemimethylated state in this region of the chromosome lasts for about one-third of the cell cycle (sequestration), but the mechanism by which it is relieved is not known. The purpose of sequestration is to prevent re-initiation from *oriC* from occurring more than once per cell cycle. For initiation to occur most efficiently, *oriC* and the *dnaA* promoter region must be fully methylated. This also contributes to ensuring that initiation occurs only once per cell cycle [9; 66]. In *Salmonella* Typhimurium, SeqA may play replication-related roles similar to those described in *Escherichia coli* [51]. In *Vibrio cholerae*, both Dam methylation and SeqA are essential [31; 54], and SeqA overproduction causes DNA replication arrest [54].

In fast-growing *Escherichia coli* or *Salmonella* Typhimurium cells, the time required for chromosome replication exceeds the doubling time. Under such conditions, *Escherichia coli* and *Salmonella* Typhimurium cells contain multiple copies of *oriC* due to initiations that occurred two or three generations ago. These origins fire simultaneously during the cell cycle, leading to synchronous initiation, which is thought to be due to the immediate release of DnaA from an origin after initiation [46]. This release will temporarily increase the DnaA / *oriC* ratio in wild-type cells for the remaining fully methylated origins. After initiation, other mechanisms ensure that DnaA is not in the proper conformation for initiation. Among these mechanisms is a reduction in the transcription of the *dnaA* gene. Sequestration by SeqA after initiation keeps the *dnaA* promoter region in a hemimethylated state, which reduces transcription initiation because the *dnaA* promoter GATC sequences need to be fully methylated for maximal expression [9].

In *dam* mutant cells, there is no sequestration by SeqA; consequently, DnaA can immediately rebind origins after the first initiation event, and initiate a second time when the concentration of the active form of DnaA is high enough. Transcription from the *dnaA* gene continues throughout the cell cycle although at a reduced level. Dam methylation, therefore, is not essential for replication initiation; rather, the cell uses methylation to discriminate between old and new origins.

7. Conclusion

In the *Enterobacteriaceae* family (*Salmonella* Typhimurium, *Escherichia coli*,…), DNA methylation and sequestration modulate a variety of processes such as DNA replication and transcription of certain genes. Deletion of the *dam* and / or *seqA* genes produces a variety of phenotypes ranging from replication asynchrony to virulence attenuation, indicating multiple functions for Dam and SeqA proteins in modulating gene expression, proper chromosome segregation, initiation of chromosome replication, and nucleic stabilization. Given these multiple roles, it is not surprising that these mutations are highly pleiotropic. However, the lack of Dam and / or SeqA proteins does not impair viability. Bacterial mutant strains are more sensitive to these mutations than the wild type which shows the inverse. In addition, no great difference between the mutants of *Salmonella* Typhimurium and those of some *Enterobacterial* species such as *Escherichia coli* was observed with replication asynchrony. In conclusion, the role of Dam and SeqA in the prokaryotic cellular processes such as the DNA replication is clear. So it may rely on their capacity as a global regulator of the gene expression during bacterial life, *in vitro*, in a similar manner as it does *in vivo*.

8. Future research

Our knowledge on the effects of Dam and SeqA proteins in *Enterobacteriaceae* family (*Salmonella* Typhimurium, *Escherichia coli*,…) has considerably improved in the last decade. This fundamental research has several implications that will prove to be useful for the development of novel therapeutic approaches. But, to date, therapeutic applications are still in their early experimental phases, but several recent studies provide promising results for future clinical developments. Over the last few years, many studies have demonstrated that *Escherichia coli* and *Salmonella typhimurium seqA* and / or *dam* mutants exhibit asynchronous DNA replication and are highly attenuated for virulence in mice and have been proposed as live vaccines. These results prove that these proteins might have a role in regulating virulence. In addition, future research must focus on the study of the decreasing virulence and the proteomic and enzymatic activities of a *seqA* and / or *dam* mutant strains. So these perspectives can be useful to more fully understand the significance of the results obtained above. Of special interests are: firstly, the growing list of genes governed by DNA methylation and sequestration in bacterial pathogens ; secondly, the finding of novel genes regulated by Dam and SeqA proteins using high throughput analysis, and, thirdly, the evidence that these proteins may regulate the expression of many unidentified genes involved in the DNA replication.

Finally, the way in which Dam and SeqA participate clearly in the DNA replication is a critical question that deserves further investigation in the near future, and may be research studies will have to identify explanations.

Acknowledgements

The work presented in this chapter was supported by the Tunisian Ministry of Higher Education, Scientific Research, and Technology. It has been performed at the biochemistry and molecular biology laboratory which belongs to the Faculty of Sciences of Bizerta. Many people have in different ways supported me in my work and contributed to the completion of this study. I would especially like to express my gratitude to:

Dr. Francisco Ramos-Morales (Departamento de Genética, Facultad de Biologia, Universidad de Sevilla, Spain).

Miss BCHINI Khouloud for proofreading and correcting this manuscript.

Author details

Amine Aloui[1*], Alya El May[1], Saloua Kouass Sahbani[1,2] and Ahmed Landoulsi[1]

*Address all correspondence to: aminealoui@yahoo.fr

1 Carthage University, Faculty of Sciences of Bizerta Zarzouna, Tunisia

2 Sherbrooke University, Faculty of Medecine Québec, Canada

References

[1] Adams, D. E., Shekhtman, E. M., Zechiedrich, E. L., Schmid, M. B., & Cozzarelli, N. R. (1992). The role of topoisomerase IV in partitioning bacterial replicons and the structure of catenated intermediates in DNA replication. *Cell*, 71, 277-88.

[2] Aloui, A., Chatti, A., May, E. L. A., & Landoulsi, A. (2007). Effect of methylation on DNA replication in Salmonella enterica serovar Typhimurium. *C R Biol*, 330, 576-580.

[3] Aloui, A., Kouass, S. S., Mihoub, M., El May, A., & Landoulsi, A. (2011). The Absence of the "GATC- Binding Protein SeqA" Affects DNA replication in Salmonella enterica serovar Typhimurium. *DNA Replication and Related Cellular Processes*, 978-9-53307-212-7.

[4] Aloui, A., Tagourti, J., May, E. L. A., Joseleau, P. D., & Landoulsi, A. (2011). The effect of methylation on some biological parameters in Salmonella enterica serovar Typhimurium. Path biol, , 59, 192-198.

[5] Bach, T., Krekling, M. A., & Skarstad, K. (2003). Excess SeqA prolongs sequestration of oriC and delays nucleoid segregation and cell division. EMBO J. , 22, 315-323.

[6] Bae, S. H., Cheong, H. K., Cheong, C., Kang, S., Hwang, D. S., & Choi, B. S. (2003). Structure and dynamics of hemimethylated GATC sites: implications for DNA-SeqA recognition. *J. Biol. Chem.*, 278, 45987 -93 .

[7] Bates, D., & Kleckner, N. (2005). Chromosome and replisome dynamics in E. coli: loss of sister cohesion triggers global chromosome movement and mediates chromosome segregation. *Cell*, 121, 899 -911 .

[8] Bogan, J. A., & Helmstetter, C. E. (1997). DNA sequestration and transcription in the oriC region of Escherichia coli. *Mol. Microbiol.*, 26, 889-896.

[9] Braun, R. E., O'Day, K., & Wright, A. (1985). Autoregulation of the DNA replication gene dnaA in E. coli K-12. *Cell*, 40, 159-169.

[10] Brendler, T., Abeles, A., & Austin, S. (1995). A protein that binds to the 1 origin core and the oriC 13mer region in a methylation-specific fashion is the product of the host seqA gene. EMBO J. , 14, 4083-4089.

[11] Brendler, T, & Austin, S. (1999). Binding of SeqA protein to DNA requires interaction between two or more complexes bound to separate hemimethylated GATC sequences. *EMBO J.*, 18, 2304-2310.

[12] Brendler, T., Sawitzke, J., Sergueev, K., & Austin, S. (2000). A case for sliding SeqA tracts at anchored replication forks during Escherichia coli chromosome replication and segregation. *EMBO J.*, 19, 6249-6258.

[13] Bergerat, A., Guschlbauer, W., & Fazakerley, G. V. (1991). Allosteric and catalytic binding of S-adenosylmethionine to Escherichia coli DNA adenine methyltransferase monitored by 3H NMR. *P Natl Acad Sci USA*, 88, 6394-7.

[14] Boye, E., & Lobner-Olesen, A. (1990). The role of dam methyltransferase in the control of DNA replication in E. coli. *Cell*, 62, 981-9.

[15] Boye, E., Marinus, M. G., & Løbner-Olesen, A. (1992). Quantitation of Dam methyltransferase in Escherichia coli. *J Bacteriol*, 174, 1682-1685.

[16] Boye, E., Stokke, T., Kleckner, N., & Skarstad, K. (1996). Coordinating DNA replication initiation with cell growth: differential roles for DnaA and SeqA proteins. *Proc. Natl Acad. Sci. USA*, 93, 12206-12211.

[17] Brendler, T., Sawitzke, J., Seerguev, K., & Austin, S. (2000). A case for sliding SeqA tracts at anchored replication forks during Escherichia coli chromosome replication and segregation. *EMBO J.*, 19, 6249-6258.

[18] Campbell, J. L., & Kleckner, N. (1990). E. coli oriC and the dnaA gene promoter are sequestered from dam methyltransferase following the passage of the chromosomal replication fork. *Cell*, 62, 967-979.

[19] Cooper, S., & Helmstetter, C. E. (1968). Chromosome replication and the division cycle of Escherichia coli B/r. J. Mol. Biol. , 31, 519-40.

[20] d'Alencon, E., Taghbalout, A., Kern, R., & Kohiyama, M. (1999). Replication cycle dependent association of SeqA to the outer membrane fraction of E. coli. *Biochimie*, 841-846.

[21] Espeli, O., Mercier, R., & Boccard, F. (2008). DNA dynamics vary according to macrodomain topography in the E. coli chromosome. Mol. Microbiol. , 68, 1418-1427.

[22] Fossum, S., Crooke, E., & Skarstad, K. (2007). Organization of sister origins and replisomes during multifork DNA replication in Escherichia coli. EMBO J. , 26, 4514-4522.

[23] Geier, G. E., & Modrich, P. (1979). Recognition sequence of the dam methylase of Escherichia coli K12 and mode of cleavage of Dpn I endonuclease. J. Biol. Chem. , 254, 1408-13.

[24] Guarné, A., Zhao, Q., Ghirlando, R., & Yang, W. (2002). Insights into negative modulation of E. coli replication initiation from the structure of SeqA-hemimethylated DNA complex. *Nat. Struct. Biol.*, 9, 839-843.

[25] Guarné, A., Brendler, T., Zhao, Q., Ghirlando, R., Austin, S., & Yang, W. (2005). Crystal structure of a SeqA-N filament: implications for DNA replication and chromosome organization. EMBO J. , 24, 1502-1511.

[26] Han, J. S., Kang, S., Lee, H., Kim, H. K., & Hwang, D. S. (2003). Sequential binding of SeqA to paired hemi-methylated GATC sequences mediates formation of higher order complexes. *J. Biol. Chem.*, 278, 34983-9.

[27] Herman, G. E., & Modrich, P. (1981). Escherichia coli K-12 clones that overproduce dam methylase are hypermutable. *J Bacteriol*, 145, 644-6.

[28] Herman, G. E., & Modrich, P. (1982). Escherichia coli dam methylase. Physical and catalytic properties of the homogeneous enzyme. *J Biol Chem*, 257, 2605-2612.

[29] Hiasa, H., & Marians, K. J. (1996). Two distinct modes of strand unlinking during theta-type DNA replication. J. Biol. Chem. , 271, 21529-35.

[30] Hiraga, S., Ichinose, C., Niki, H., & Yamazoe, M. (1998). Cell cycle-dependent duplication and bidirectional migration of SeqA-associated DNA-protein complexes in E. coli. *Mol. Cell*, 1, 381-7.

[31] Julio, S. M., Heithoff, D. M., Provenzano, D., Klose, K. E., Sinsheimer, R. L., Low, D. A., & Mahan, M. J. (2001). DNA adenine methylase is essential for viability and plays a role in the pathogenesis of Yersinia pseudotuberculosis and Vibrio cholerae. *Infect Immun*, 69, 7610-5.

[32] Joo, S. H., Sukhyun, K., Sung, H. K., Min, J. K., & Deog, S. H. (2004). Binding of SeqA protein to hemi-methylated GATC sequences enhances their interaction and aggregation properties. J. Biol. Chem. Issue of July 16 , 279(29), 30236-30243.

[33] Kang, S., Han, J. S., Park, J. H., Skarstad, K., & Hwang, D. S. (2003). SeqA protein stimulates the relaxing and decatenating activities of topoisomerase IV. J. Biol. Chem. , 278, 48779-85.

[34] Kang, S., Lee, H., Han, J. S., & Hwang, D. S. (1999). Interaction of SeqA and Dam methylase on the hemimethylated origin of Escherichia coli chromosomal DNA replication. J. Biol. Chem., 274, 11463-8.

[35] Kato, J., Nishimura, Y., Yamada, M., Suzuki, H., & Hirota, Y. (1988). Gene organization in the region containing a new gene involved in chromosome partition in Escherichia coli. J. Bacteriol., 170, 3967-3977.

[36] Kato, J., Nishimura, Y., Imamura, R., Niki, H., Hiraga, S., & Suzuki, H. (1990). New topoisomerase essential for chromosome segregation in E. coli. Cell, 63, 393-404.

[37] Klungsøyr, H. K., & Skarstad, K. (2004). Positive supercoiling is generated in the presence of Escherichia coli SeqA protein. Mol. Microbiol., 54, 123-131.

[38] Kornberg, A., & Baker, T. A. (1992). DNA Replication, nd Ed. 521-524, W. H. Freeman and Co., New York.

[39] Landoulsi, A., Malki, A., Kern, R., Kohiyama, M., & Hughes, P. (1990). The E. coli cell surface specifically prevents the initiation of DNA replication at oriC on hemimethylated DNA templates. Cell, 63, 1053-1060.

[40] Løbner-Olesen, A., Marinus, M. G., & Hanssen, F. G. (2003). Role of SeqA and Dam in Escherichia coli gene expression: a global/microarray analysis. Proc. Natl. Acad. Sci. USA., 100, 4672-4677.

[41] Lu, M., Campbell, J. L., Boye, E., & Kleckner, N. (1994). SeqA: a negative modulator of replication initiation in E.coli. Cell, 413-426.

[42] Marinus, M. G., Poteete, A., & Arraj, J. A. (1984). Correlation of DNA adenine methylase activity with spontaneous mutability in Escherichia coli K-12. Gene, 123 EOF-5 EOF.

[43] Messer, W., & Weigel, C. (1996). Escherichia coli and Salmonella typhimurium: Cellular and Molecular Biology. Neidhardt, F. C., ed) 2nd Ed., 1580-1582, American Society for Microbiology, Washington, D. C.

[44] Morigen, , Odsbu, I., & Skarstad, K. (2009). Growth rate dependent numbers of SeqA structures organize the multiple replication forks in rapidly growing Escherichia coli. Genes Cells, 14, 643-57.

[45] Nielsen, H. J., Li, Y., Youngren, B., Hansen, F. G., & Austin, S. (2006). Progressive segregation of the Escherichia coli chromosome. Mol. Microbiol. , 61, 383-393.

[46] Nielsen, O., & Løbner-Olesen, A. (2008). Once in a lifetime: strategies for preventing re-replication in prokaryotic and eukaryotic cells. EMBO Rep, 9, 151-156.

[47] Odsbu, I., Klungsoyr, H. K., Fossum, S., & Skarstad, K. (2005). Specific N-terminal interactions of the Escherichia coli SeqA protein are required to form multimers that restrain negative supercoils and form foci. Genes Cells, 10, 1039-1049.

[48] Ogden, G. B., Pratt, M. J., & Schaechter, M. (1988). The replicative origin of the E. coli chromosome binds to cell membranes only when hemimethylated. *Cell*, 54, 127-35.

[49] Palmer, B. R., & Marinus, M. G. (1994). The dam and dcm strains of Escherichia coli-a review. *Gene*, 143, 1-12.

[50] Peterson, S. N., & Reich, N. O. (2006). GATC flanking sequences regulate Dam activity: evidence for how Dam specificity may influence pap expression. *J Mol Biol*, 355, 459-472.

[51] Prieto, A. I., Jakomin, M., Segura, I., Pucciarelli, M. G., Ramos-Morales, F., del Portillo, F. G., & Casadesus, J. (2007). The GATC-binding protein SeqA is required for bile resistance and virulence in Salmonella enterica Serovar Typhimurium. J. Bacteriol. , 189, 8496-8502.

[52] Reyes-Lamothe, R., Possoz, C., Danilova, O., & Sherratt, D. J. (2008). Independent positioning and action of Escherichia coli replisomes in live cells. *Cell*, 133, 90-102.

[53] Ringquist, S., & Smith, C. L. (1992). The Escherichia coli chromosome contains specific, unmethylated dam and dcm sites. P Natl Acad Sci USA, , 89, 4539-4543.

[54] Saint-Dic, D., Kehrl, J., Frushour, B., & Kahng, L. S. (2008). Excess SeqA leads to replication arrest and a cell division defect in Vibrio cholerae. *J Bacteriol*, 190, 5870-5878.

[55] Skarstad, K., Boye, E., & Steen, H. B. (1986). Timing of initiation of chromosome replication in individual Escherichia coli cells. EMBO J. , 5, 1711-1717.

[56] Skarstad, K., von, Meyenburg. K., Hansen, F. G., & Boye, E. (1988). Coordination of chromosome replication initiation in Escherichia coli: effects of different dnaA alleles. J. Bacteriol. , 170, 852-858.

[57] Skarstad, K., & Lobner-Olesen, A. (2003). Stable co-existence of separate replicons in Escherichia coli is dependent on once-per-cell-cycle initiation. EMBO J. , 22, 140-50.

[58] Slater, S., Wold, S., Lu, M., Boye, E., Skarstad, K., & Kleckner, N. (1995). E. coli SeqA protein binds oriC in two different methyl-modulated reactions appropriate to its roles in DNA replication initiation and origin sequestration. *Cell*, 82, 927-936.

[59] Slominska, M., Wegrzyn, A., Konopa, G., Skarstad, K., & Wegrzyn, G. (2001). SeqA, the Escherichia coli origin sequestration protein, is also a specific transcription factor. Mol. Microbiol. , 40, 1371-1379.

[60] Urig, S., Gowher, H., Hermann, A., Beck, C., Fatemi, M., Humeny, A., & Jeltsch, A. (2002). The Escherichia coli dam DNA methyltransferase modifies DNA in a highly processive reaction. *J Mol Biol*, 319, 1085-1096.

[61] Waldminghaus, T., & Skarstad, K. (2009). The Escherichia coli SeqA protein. *Plasmid*, 61, 141-150.

[62] Wegrzyn, A., Wrobel, B., & Wegrzyn, G. (1999). Altered biological properties of cell membranes in Escherichia coli dnaA and seqA mutants. *Mol. Gen. Genet.*, 261, 762-769.

[63] Weitao, T., Nordstrom, K., & Dasgupta, S. (1999). Mutual suppression of mukB and seqA phenotypes might arise from their opposing influences on the Escherichia coli nucleoid structure. Mol. Microbiol. , 34, 157-168.

[64] Weitao, T., Nordstrom, K., & Dasgupta, S. (2000). Escherichia coli cell cycle control genes affect chromosome superhelicity. EMBO Rep. , 1, 494-499.

[65] Wion, D., & Casadesus, J. (2006). N6-methyl-adenine: an epigenetic signal for DNA-protein interactions. Nat. Rev. Microbiol. , 4, 183-192.

[66] Yamaki, H., Ohtsubo, E., Nagai, K., & Maeda, Y. (1988). The oriC unwinding by dam methylation in Escherichia coli. *Nucleic Acids Res*, 16, 5067-5073.

[67] Yamazoe, M. S., Adachi, S., Kanava, K. O., & Hiraga, S. (2005). Sequential binding of SeqA protein to nascent DNA segments at replication forks in synchronized cultures of E. coli. *Mol. Microbiol.*, 55, 289-298.

[68] Zechiedrich, E. L., & Cozzarelli, N. R. (1995). Roles of topoisomerase IV and DNA gyrase in DNA unlinking during replication in Escherichia coli. *Genes Dev*, 9, 2859-69.

Chapter 5

Chromatin Damage Patterns Shift According to Eu/ Heterochromatin Replication

María Vittoria Di Tomaso, Pablo Liddle,
Laura Lafon-Hughes, Ana Laura Reyes-Ábalos and
Gustavo Folle

Additional information is available at the end of the chapter

1. Introduction

In order to maintain genetic stability, strictly controlled mechanisms are essential to assure the accuracy of genetic functions. Precise genome replication and correct control of gene expression mostly *via* epigenetic mechanisms are critical in maintaining the stability of genomes. Moreover, the characteristic chromatin compartmentalization of mammalian genomes contributes to regulate the housekeeping or tissue-specific genetic activities [1, 2].

Table 1 summarizes the distinct chromatin compartments and their foremost properties. Euchromatin (*eu*: true) and heterochromatin (*hetero*: different) are two major compartments or chromatin states of the DNA originally distinguished by their isopycnotic or heteropycnotic interphase staining properties, respectively [3]. The heterochromatin compartment differentiates in both constitutive (permanent) and facultative (developmentally reorganized) states [4]. Facultative heterochromatin represents chromatin regions being facultatively inactivated (heterochromatinized) because of gene dosage compensation (i.e.: mammalian female inactive X chromosome) randomly silenced at an early stage of embryogenesis or tissue-specific gene expression. Constitutive heterochromatin consists in regions of α- and β-heterochromatin [5, 6].

Distinct features characterize the different chromatin states (Table 1). Interphase open chromatin conformation and transcriptional activity in all cell types distinguish euchromatin. Higher order chromatin compaction characterizes constitutive α- and β-heterochromatic regions while gene silencing differentiates constitutive α-heterochromatin. Tissue-specific transcriptional activity and low or high chromatin condensation, depending on gene expression, correspond to features of facultative heterochromatin [7, 6]. The mammalian genome compartmentalization can be visualized in both banded metaphase chromosomes and stained interphase nuclei.

Compartments	Euchromatin	Facultative heterochromatin		Constitutive heterochromatin	
Chromatin types	Euchromatin	Tissue-specific	Dosage compensation	α- heterochromatin	β-heterochromatin
Location in metaphase chromosomes	Light G-bands	Dark G-bands	Inactive X chromosome (Xi)	C-bands	C-bands
Location in interphase nuclei	Inner compartment	Peripheral compartment and chromocenters	Peripheral compartment and chromocenters	Peripheral compartment and chromocenters	Peripheral compartment and chromocenters
Interphase chromatin compaction	Open conformation	Low or high order compaction	High order compaction	High order compaction	High order compaction
Presence of genes	Housekeeping genes	Inactivated tissue-specific genes	Dosage inactivated genes	No genes	Transposable elements and heterochromatic genes
Gene expression and relation to chromatin state	Gene activity in euchromatic state in all cells	Tissue-specific gene activity in euchromatic state	Gene activity in euchromatic state until silencing	No gene activity	Low gene activity in heterochromatic state
GC or AT DNA sequences richness	GC-rich	AT-rich	GC- and AT-rich	AT-rich	AT-rich
Repeated DNA sequences	SINEs	LINEs	SINEs and LINEs	Tandem highly repeated DNA sequences	Tandem highly repeated DNA sequences
CpG island methylation	Unmethylated	Unmethylated or methylated	Methylated	Methylated	Methylated
Core histone tail acetylation	Hyperacetylated	Hyperacetilated or hypoacetylated	Hypoacetylated	Hypoacetylated	Hypoacetylated
Replication timing	Early	Early or late	Late	Late	Late or early

Table 1. Distinguishing properties of chromatin compartments.

The C-banding procedure [8] produces a selective staining of specific chromosome regions, mapping at or adjacent to centromeres, telomeres or interstitial arm sites, depending on the species. Occasionally, a chromosome arm is entirely heterochromatic, such as the long arm of the Chinese hamster X chromosome (Figure 1, left). In humans, C-bands are located at centromeres and pericentric regions of all chromosomes, being conspicuous at the pericentric regions of chromosomes 1, 9 and 16 and the distal long arm of the Y chromosome (Yq) (Figure 1, right).

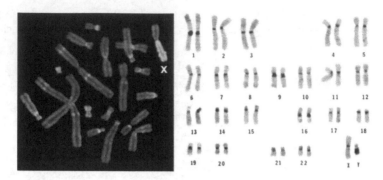

Figure 1. C-banding in CHO and human chromosomes. Left: C-banded metaphase of CHO9 cell line. The CHO cell line was established from a Chinese hamster ovary fibroblast culture [9] and presents a modal number of 21 chromosomes. This cell line contains eight normal and twelve rearranged autosomes with only one X chromosome. Giemsa-stained C-band regions are visualized in yellow (reflected light microscopy). The CHO X chromosome (X) shows an almost entirely heterochromatic long arm. Right: C-banded caryotype of a human peripheral lymphocyte metaphase showing centromeric, pericentric (chromosomes 1, 9 and 16) and distal Yq heterochromatic blocks.

By digestion with the proteolytic enzyme trypsin followed by Giemsa staining (G-banding procedure) [10], a pattern of alternate light and dark regions along the length of all chromosomes is obtained (light G-bands and dark G-bands, respectively). The G-band pattern is characteristic for each chromosome pair allowing their precise identification and caryotyping. Figure 2 shows the CHO9 and human G-band chromosome patterns.

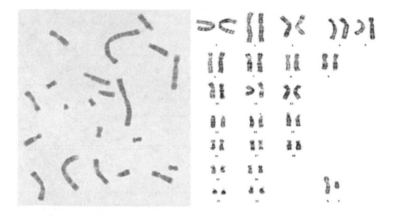

Figure 2. G-banded CHO9 metaphase (left) and a male human peripheral lymphocyte caryotype exhibiting G-bands (right).

C- and G-band patterns reveal the heterogeneous organization of chromatin along condensed chromosomes. C-bands enclose constitutive α- and β-heterochromatin. Regions with ubiquitously expressed housekeeping genes (euchromatin) reside in light G-bands, while tissue-specific genes (facultative heterochromatin) dwell in dark G-bands [5, 6, 11].

Light and dark G-bands may reflect a differential array of SAR (Scaffold-Associated Regions), composed by highly AT-rich DNA stretches binding to the chromosome scaffold. Regions of dark G-bands exhibit a tighter chromatin fiber coiling than light G-bands domains [12]. Constitutive heterochromatin has an even more dense conformation.

Moreover, euchromatic light G-bands are GC-rich and gene-dense regions, containing unmethylated CpG islands and moderately repeated Short Interspersed Elements (SINE), mainly represented by Alu family sequences. Conversely, facultative heterochromatic dark G-bands are AT-rich, gene-poor and harbor hypermethylated CpG and moderately repeated family of Long Interspersed Elements (LINE) sequences. Constitutive α-heterochromatic C-bands are the major locations of tandem non-coding highly repeated satellite DNA sequences, devoid of genes [11, 13]. However, constitutive β-heterochromatin presents inserted middle-repetitive transposable elements between the tandem repeats, some of them transcriptionally active [6]. Moreover, genes residing within regions of pericentric constitutive β-heterochromatin termed "heterochromatic genes" have been reported in *Drosophila*, mammals and plants [14, 15].

In spite of variations according to cell type or function of mammalian interphase nuclei, the corresponding chromatin of light and dark G-bands as well as C-bands is non-randomly distributed in different nuclear compartments, displaying specific chromatin conformation, molecular composition and gene expression patterns.

In most interphase cells, euchromatin (light G-bands) dwells in the inner compartment of nuclei, whereas heterochromatin (dark G-bands and C-bands) resides in the peripheral compartment, chromocenters and around nucleoli [6, 16]. Figure 3 illustrates a HeLa nucleus where the different interphase chromatin compartments can be recognized.

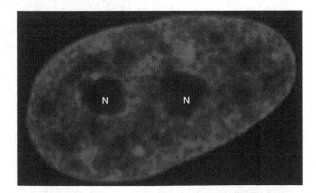

Figure 3. Distinct eu/heterochromatin compartments in DAPI-stained HeLa interphase nucleus. DAPI-bright regions correspond to heterochromatin and dim areas to euchromatin. N: nucleolus.

Constitutive and tissue-specific genes are only expressed in the euchromatic state. Therefore, facultative heterochromatin behaves as euchromatin in cells where its tissue-specific genes are transcribed, but holds a packed (heterochromatic) conformation when genes remain silent.

However, some transposons and heterochromatic genes of β-heterochromatin are transcriptionally active in heterochromatic state suggesting that distinct epigenetic mechanisms of gene regulation and preservation of eu/heterochromatic states may exist in these regions [6, 14, 15].

Once acquired, the chromatin states are somatically maintained as stable heritable epigenetic states. Euchromatin remodels during mitosis and restores the original organization in early G_1 phase of each cell cycle. In addition, during DNA synthesis (S-phase) both euchromatin and heterocromatin transiently lose their typical condensation status recovering the previous folding level after replication. Establishment and maintenance of chromatin states involve post-translational modification enzymes that act coordinately to methylate CpG islands and to either acetylate, methylate, phosphorylate, ubiquitinate, poly-ADP ribosylate or SUMOylate the core histone tails of nucleosomes. These epigenetic changes, together with the recruitment of methyl-CpG binding proteins, ATP-dependent chromatin remodeling complexes and the association of specific non-histone proteins, such as HP1 (Heterochromatin Protein 1) or RNAi (non-coding interference RNA), also mediate the regulation of DNA replication, transcription and repair [17, 18].

The N- and C-termini of H3 and H4 core histones are particularly involved in epigenetic regulation. Acetyl groups covalently added to lysines, serines or arginines of the N-terminal histone tails reduce the affinity to DNA, promoting the accessibility of chromatin remodeling and activating transcription factors. Therefore, histone hyperacetylation usually characterizes active chromatin regions. Conversely, transcriptionally silenced chromatin regions generally contain hypoacetylated histones (Table 1). For instance, H3 acetylated (ac) in lysine 9 (H3K9ac) is enriched at the promoter region of active genes although, it was reported that the histone H3 acetylated at lysine 4 (H3K4ac) resides in pericentric heterochromatin of *Schizosaccharomyces pombe*, playing a role in the assembly of repressive heterochromatin [19]. On the other hand, histone methylation (me) can be associated with transcriptional activation or repression. For example, methylation of H3 on lysines 4, 36 or 79 (H3K4me, H3K36me, H3K79me) is associated with transcriptional activation whereas methylation of H3 on lysines 9 or 27 (H3K9me, H3K27me) and of H4 on lysine 20 (H4K20me) is involved in transcriptional repression [18]. The concerted action of acetylated and methylated histone core residues is central in creating a "histone code" which delineates distinct genomic loci that recruit factors needed for DNA remodeling, transcription, replication and repair [5, 17].

In general, methylation of CpG islands within 5'regions of genes is associated with hypoacethylated histones, characterizing the heterochromatic state (Table 1). However, DNA methylation is not exclusively related to gene silencing. It was reported that methylation of some imprinting centers can displace trans-acting repressor factors, allowing the expression of the linked imprinted genes [20].

The epigenetic mechanisms involved in the maintenance of eu/heterochromatic compartments and gene expression are connected to DNA replication. There are specific interactions between components of the replication machinery and chromatin related factors, timing the eu- or heterochromatin replication.

2. Replication of eu/heterochromatin compartments

Compartmentalization of vertebrate genomes cooperates in achieving the high fidelity DNA replication necessary for the accurate preservation of the genetic information throughout cell generations. DNA replication is a temporarily and spatially highly ordered and strictly regulated process, occurring during S-phase of the cell cycle, with distinct genome compartments replicating at different times. The replication timing of the genome compartments are highly conserved within consecutive cell cycles and regulated by specific epigenetic chromatin conformation domains, DNA features and transcriptional activity [21, 22, 23].

Mammalian chromosome duplication involves clusters or domains of neighboring replicons named Replication Timing Domains (RTD) which synchronously start and end replication, according to a deterministic replication timing program [21, 22, 24]. When one domain completes replication, an adjacent domain successively initiates DNA synthesis [25]. Remarkably, mouse and human asynchronous replication timing may function randomly between individual replicons within a RTD and non-randomly between RTD [25]. The random firing of replication origins within a RTD generates a different replication pattern during each S-phase, but it has been reported that some origins fire preferentially and more frequently than others [26]. The RTD are stable structures of mammalian interphase nuclei, replicating and transcribing in temporal and spatial coordination [26].

Pulse labeled interphase nuclei of human, mouse and hamster cells with the base analogues 5-bromo-2'-deoxyuridine (BrdU) or 5-ethynyl-2'-deoxiuridine (EdU) demonstrated the asynchrony and specific spatial distribution of DNA replication. The early replication pattern of S-phase (ES-phase) is characterized by replication foci dispersed throughout the inner environment of the nuclei with scarce or absence of foci at the periphery or adjacent to the nucleoli. The replication pattern changes throughout the progression of S-phase. In mid S-phase (MS-phase) most foci map adjacent to the internal nuclear membrane and around nucleoli, with few foci centrally located. Lastly, late S-phase replication maps next to the nuclear envelope as well as in chromocenters and around nucleoli [16, 27]. Early S-phase and late S-phase replication patterns of CHO9 cells are illustrated in Figure 4.

In general, chromatin with transcriptional activity (euchromatin) replicates early in S-phase whereas constitutive α-heterochromatin duplicates late. Besides, facultative heterochromatin replicates earlier if its tissue-specific genes are being expressed and later if not [6, 28] (Table 1). It has been reported that genes of mouse embryonic stem cells residing within GC-rich and LINE-poor DNA (euchromatin) do not modify their replication timing after differentiation to neural precursors, whereas genes residing in AT-rich and LINE-rich DNA revealed changes in replication timing accompanied by changes in gene expression and chromatin folding [29]. A change of replication timing from early S to late S-phase is particularly evident in the female mammalian Xi [30]

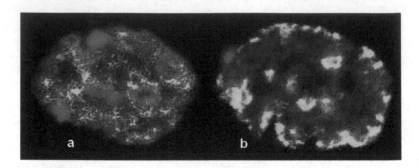

Figure 4. Early (ES-phase) or late (LS-phase) replication patterns of CHO9 nuclei revealed by incorporation of EdU and subsequent detection with an Alexa Fluor 488 (green) conjugated azide (Click-iT EdU imaging kit, Invitrogen). (a) ES-phase nucleus with inner compartment replication. (b) LS-phase nucleus showing replication in the peripheral compartment, chromocenters and around nucleoli.

Early replication seems to be important but not essential for gene transcription. Moreover, late replication is not an obligatory feature of heterochromatin. For example, transcriptionally active transposons of β-heterochromatin replicate late while the heterochromatic centromeres and the silent mating-type cassettes of *Schizosaccharomyces pombe* replicate in early S phase [14, 15, 31]. There are additional cases reported of early heterochromatin replication such as human telomeres [32]. and mouse pericentric heterochromatin and centromeres [33].

The early replicon clusters of higher eukaryotes alternate their replication and transcription activity. However, correlation between replication and transcription does not exist in *Saccharomyces cerevisiae* [34]. Employing distinct colored fluorescent labels to recognize early replication foci and transcription foci (factories), it was shown that both labels do not colocalize. In ES-phase, actively replicating foci are transcriptionally inactive and only restart transcription after finishing replication. The replication timing is indirectly related to transcription through the assembly of a higher-order chromatin state [2]. For example, silencing of the mammalian Xi is initially reversible and only stabilizes when an identifiable higher-order chromatin configuration (Barr body) appears and replication is delayed [35].

The chromatin replication timing is reestablished early in G_1 phase of each cell cycle, coincident with the anchorage and positioning of chromosomal segments at specific locations within the nucleus named TDP (**T**iming **D**ecision **P**oint) [36]. Both anchorage and positioning of chromosomes are central in the organization of nuclear eu/heterochromatic compartments and the establishment of replication timing and transcriptional activity [23, 36]. Modifications in subnuclear chromatin organization are associated with changes in replication timing during development [37]. For example, the position of the immunoglobulin heavy chain locus (IgH) in B cells shows that its localization in the interphase nuclei depends on replication timing and gene activity. During early stages of B cell differentiation, both transcribed alleles of the IgH locus are centrally located in the nucleus and replicate early. Conversely, in advanced differentiation stages the IgH locus is repositioned to the nuclear periphery, repressed and late-replicated [38].

Nonetheless, chromatin positioning at the nuclear periphery is indicative but not mandatory for gene silencing and late replication. In fact, the nuclear periphery is heterogeneous with respect to transcription. For instance, in budding yeast, nuclear pores, which mediate the transport between the nucleus and cytoplasm, enhance the transcriptional activity of genes positioned in their proximity [39]. The dosage compensation complex of the hyperacetylated *Drosophila* male X chromosome interacts with nuclear pore proteins determining its transcription up-regulation and early DNA duplication [40].

Replication clusters correspond to bands of metaphase chromosomes. Tightly coiled C-band (constitutive heterochromatin) replicates in late S-phase. Facultative heterochromatin of the dark G-bands duplicates either early or late depending on its tissue-specific expression. Early replication pattern characterizes the loosely coiled euchromatin of light G-bands. Ubiquitously expressed housekeeping genes (light G-bands) are therefore early replicating [41, 42]. Duplication timing analysis by quantitative PCR of the boundary region between G-light 13q14.3 and G-dark 13q21.1 bands showed that the G-light side of the frontier replicates early whilst the G-dark interface replicates late. However, analysis using PCR primers spaced at approximately 150 Kb intervals showed that the switch in G-light/G-dark band replication timing takes place gradually from early-mid to late S-phase over a 1-2 Mb region [43]. The DNA segments corresponding to large regions between early and late-S phase replication timing domains are termed TTR (Timing Transition Regions) [44].

A correlation between replication timing and epigenetic modification of chromatin has also been shown. Early replication domains are related to specific combination of changes in histone lysine residues (H3K9Ac, H3K27Ac, H3K4me, H3K36me and H3K79me) associated with transcriptional activity. On the other hand, the repressive epigenetic modifications (H3K9me, H3K27me and H4K20me) are linked to late replication [18].

Chromatin epigenetic changes occurring throughout DNA replication may provide a replication timing mechanism (firing early or late replication origins) in the direction of maintaining specific chromatin expression patterns [45]. It was reported that histone hypoacetylation is needed to preserve normal heterochromatin replication dynamics [46] and that histone hyperacetylation may increase the efficiency of replication origins, advancing the replication timing of distinct genomic regions [47]. For instance, removal of acetyl groups by HDAC (**H**istone **DeAC**etylase) contributes to mantain late replication at imprinted loci [48] and the generation of neocentromeres [49].

Several proteins, including CpG island-methylating DNMT (**DNA M**ethyl **T**ransferase), core histone tail-methylating HMT (**H**istone **M**ethyl **T**ransferase) and HP1 (**H**eterochromatin-associated **P**rotein), colocalize with late replicating DNA regions [45]. HP1 binds to heterochromatin, facilitating the extension of the repressive H3K9me modification [50] and hence delaying replication timing by supporting heterochromatin conformation. HP1 could facilitate the late firing of replication origins within heterochromatin [51]. Furthermore, mutations of DNMT result in earlier replication of normally late replicating DNA. For instance, patients with mutations in the Dnmt3b gene (coding protein DNMT3b) have hypomethylated CpG islands in the Xi chromosome, which replicates at an earlier S-phase stage despite the presence of XIST (X-Inactive Specific Transcript) RNA [52]. Accordingly, changes in either DNA or histone methylation status in concert with histone acetylation patterns may promote open or tight chromatin conformations and thus modifications in the firing of replication origins and/or replication rates [47].

In mammals, several distinct discrete or diffuse genomic sequence motifs can potentially act as Origin Replication Identification (ORI), where a large number of proteins bind to load replication complexes. A protein complex, named the pre-Replication Complex (pre-RC) associates with potential replication origins in G_1 phase. This complex includes the Origin Recognition Complex (ORC), which recognizes the replication origins, the helicase MCM2-7 (Mini Chromosome Maintenance 2-7), and other essential factors. Early firing ORI demonstrated to be rich in MCM proteins. Besides, MCM could be more efficient in early firing than in late firing ORI suggesting that heterochromatin could repress MCM activities [53, 54].

Accessibility of replication initiation factors to redundant or discrete replication origins may be regulated by its nuclear localization in relation to chromatin states. For example, the early replicating α-globin locus is located within a light G-band. Deletions that juxtapose the α-globin locus next to a region of late replicating telomeric condensed heterochromatin (repositioning this locus to the nuclear periphery), delay the initiation of α-globin replication by restricting the access of initiation factors to the ORI [55].

There is a complex cell cycle intra-S checkpoint involving the ATR/CHK1-related network in metazoas and ATR/Rad53 in *Saccharomyces cerevisiae* that controls replication asynchrony. The transition from early to late S-phase replication (mid-S replication pause) is coupled with the activation of the intra S-phase checkpoint at mid S-phase which inhibits the initiation of late replicons. It has been reported that inhibition of CHK1 generates earlier firing of a late-firing subset of ORI [56, 57]. Accordingly, the checkpoint function may play a role in regulating replication asynchrony and S-phase progression [25, 58].

Both DNA and histone methylation can affect replication timing *via* the ATR/CHK1 control pathway. There is a complex and so far not completely understood relationship between checkpoint function and epigenetic modifications (DNA methylation, histone methylation and histone acetylation) in the regulation of replication origins firing during S-phase [47, 59].

Following pre-RC loading to ORI, a protein pre-Initiation Complex (pre-IC) assembles upon MCM proteins together with factors required for loading replicative polymerase. The chromatin association of pre-RC and pre-IC is asynchronous, allowing pre-RC inhibition and pre-IC activation (from S-phase initiation toward the end of mitosis) by the cell cycle CDK proteins (Cyclin-Dependent Kinases). This regulation licenses replication to occur at a specific time, only once per cell cycle, and ensures that cell cycle cannot progress until checkpoints are satisfied. In *Xenopus laevis* and mammalian cells there is an additional system to control licensing by means of the geminin protein, which also inhibits pre-RC. Degradation of geminin at the end of mitosis is essential for a new license of replication [56, 60].

Completion of replication is necessary for entire chromosome condensation. *Drosophila* ORC mutants unable to complete S-phase have defects not only in DNA replication (with some euchromatic regions replicating even later than heterochromatin) but also in cell cycle progression and chromatin condensation [61]. Although some levels of chromosome condensation occur in the absence of a complete replication cycle, mitotic chromosomes are shorter and thicker than in wild type *Drosophila*. Even though ORC is principally involved in the initiation of DNA replication, additional roles in mitotic chromosome condensation, centromere function as well as the establishment and maintenance of gene silencing and heterochromatin have been suggested [61, 62, 63].

3. Eu/heterochromatin replication and distribution of genetic damage

The S-phase of the cell cycle has proved to be very sensitive to genetic damage. S-phase has been considered as one of the sources of genomic instability. There are several lines of evidence that correlate genomic instability with chromosomal aberrations (CA), birth defects and infertility [64]. Besides, oncogene activation or tumor suppressor gene repression can arise as a consequence of primary DNA damage or CA [65]. Several authors have reported the colocalization of induced CA breakpoints (BP) (sites of chromosomal breaks in a CA) with regions harboring fragile sites, oncogenes or cancer-associated CA [66-72].

The human genome holds long stretches of AT-rich sequences as well as inverted, mirror or direct tandem repeats, prone to be arranged in unusual DNA secondary structures that may inhibit replication. The presence of secondary structures, unstable single-stranded or non-replicated regions could lead to chromosome fragility expressed as gaps or breaks in metaphase chromosomes [73, 74].

DNA replication in mammals slows down significantly when the 1-2 Mb regions of TTR are replicated [57]. It was reported that after replication of euchromatic light G-bands, the replication fork stalls at TTR of the interband regions, restarting DNA synthesis at the adjacent dark G-band after a mid S-phase pause [6]. This interband region devoid of replication origins is often replicated by means of a single replication fork [75]. Such genomic segments could generate damage-prone regions that frequently overlap with DNA fragile sites [43, 76]. For example, the common fragile site FRA3B is devoid of replication origins and thus completes replication very late in S-phase [77]. In addition, it was observed that mutation rates increase with the distance from replication origins [78, 79].

Furthermore, it was reported an increase in mutation rate as S-phase advances. Early replicating housekeeping genes are more conserved than later replicating tissue-specific genes [57, 80]. Genes corresponding to mutational hot spots involved in speciation and adaptive radiation response are late replicating [57]. CpG methylation status of late replicating regions may contribute to the rise in mutation rate mostly due to $5^{me}CpG$ substitutions [81, 82].

3.1. Eu/heterochromatin replication and induced-damage distribution in a mitotic chromosome model

DNA lesions trigger a DNA Damage Response (DDR) characterized by activation of cell cycle checkpoints, damage sensor proteins, DNA repair mechanisms and apoptotic pathways [83, 84]. The DNA Double-Strand Break (DSB) is the critical DNA lesion involved in CA production [85]. DSB can be generated by DNA-damaging agents or spontaneously through the endogenous production of reactive oxygen species (ROS) or cellular processes such as DNA replication, repair, transposition or mitotic recombination. Agents inducing DSB and CA are named clastogens. The S-phase independent clastogens, like ionizing radiation and the radiomimetic agent bleomycin, directly induce DSB. Conversely, S-phase dependent clastogens such as UV-C and alkylating compounds need the intervention of DNA repair and replication in order to generate DSB, which could ultimately lead to CA. Hence, DNA replication constitutes a relevant step in the transformation of DNA lesions into CA. Besides, some clastogenic agents such as the anti-topoisomerase II cleavable complex trappers behave as S-phase independent clastogens. Eukaryotic topoisomerases II alleviate tensional DNA stress by the generation of a DNA topoisomerase II complex (cleavable complex) within which the topoisomerase II component introduces transient breaks in both DNA strands (DSB) allowing the DNA to pass through the breaks [86]. Drugs that act by trapping cleavable complexes hamper the resealing of DSB produced by topoisomerase II and, as a consequence, DNA DSB persist [87, 88].

As shown in Figure 1, the CHO9 X-chromosome exhibits an almost entire constitutive heterochromatic long arm (Xq) with the exception of a medial secondary constriction. Besides, Xq replicates in late S-phase whereas the euchromatin of the short arm (Xp) and the Xq secondary constriction duplicates during early S-phase (Figure 5) [89, 90]. Differential replication timing of Xp and Xq of CHO cells provided a valuable experimental model to analyze the relationship between eu/heterochromatin DNA replication and CA induced by different types of clastogens: UV-C light, the methylating agent methylmethane sulphonate (MMS) and the anti-topoisomerase II inhibitor etoposide (a cleavable complex trapper) in BrdU pulse-labeled CHO9 chromosomes [91, 92].

CHO9 cells were treated with MMS (20 mM) or etoposide (20 μM) and simultaneously exposed to 30 mM BrdU (40 min) or otherwise exposed to UV-C (30 J/m2; 0.1 J/m2/s) and immediately labeled with BrdU (40 min). Incorporation of BrdU in Xp or Xq was disclosed by immunolabeling either treated or control CHO9 metaphases with anti-BrdU antibodies coupled to FITC. The relationship between replication timing, chromatin conformation and genetic damage was investigated by mapping induced BP in Xp and Xq in cells treated both in early and late S-phase [91, 92].

Examples of CA induced by MMS, etoposide and UV-C in replicating CHO9 Xp or Xq are shown in Figure 5. Figure 6 illustrates Xp/Xq distribution of etoposide, UV-C and MMS-induced BP in relation to replication.

The application of χ^2 test to analyze the association between Xp/Xq replication pattern and Xp/Xq BP localization showed that when Xp replicates, BP produced by either MMS, UV-C or etoposide clustered in Xp. On the other hand, during Xq replication, BP induced by the clastogens concentrated in Xq [91, 92] (Figure 6).

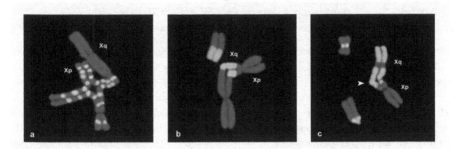

Figure 5. Illustrations of CA involving CHO9 Xp or Xq induced by (a) MMS, (b) etoposide, or (c) UV-C in (a) early (Xp replication) or (b and c) late (Xq replication) S-phase. Different types of CA are shown: (a) symmetric quadrirradial affecting Xp; (b) asymmetric quadrirradial with acentric fragment involving Xq; (c) duplication-deletion in Xq (arrow). Chromosomes exhibit BrdU immunolabeling (yellow) and either PI (red) or DAPI (blue) counterstaining.

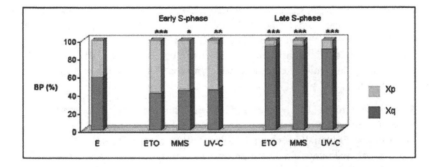

Figure 6. Bar diagram illustrating CHO9 X chromosome BP distribution induced by etoposide (ETO, 20 µM), methylmethane sulphonate (MMS, 20 mM) and UV-C (30 J/m^2; 0.1 J/m^2/s) in Xp (grey) and Xq (blue) during early (ES-phase) and late (LS-phase) cell cycle phases. The bar on the left side (E) indicates the expected frequencies of induced BP according to Xp and Xq relative length.

Since UV-C and MMS are S-phase dependent clastogens, the observed predominance of BP produced in Xp or Xq according to replication timing could be explained based on their requirement of DNA synthesis to produce CA. DNA base damage induced by MMS as well as cyclobutane pyrimidine dimers (CPD) and 6-4 photoproducts (6-4 PP) produced by UV-C are preferentially repaired through **B**ase **E**xcision **R**epair (BER) and **N**ucleotide **E**xcision **R**epair (NER) mechanisms, respectively. Both repair systems create an excision repair **S**ingle-**S**trand **B**reak (SSB) intermediate at the site of DNA lesion which is then filled by DNA repair synthesis [93]. If DNA replication initiates with an excision repair SSB intermediate, another SSB can be generated in the complementary DNA strand, thus forming a DSB [94, 95]. Additionally, CPD, 6-4 PP or base damage in a single strand (unrepaired before DNA replication) may stall the replication fork and as a result, may produce a SSB in the opposite DNA strand [96, 97]. Furthermore, two nearby SSB in each DNA strand may behave as a DSB [98]. The DSB generated could be ultimately processed and transformed in CA [91, 92].

Nonetheless, the preferential location of CA in replicating Xp or Xq during etoposide treatment (independently of its eu/heterochromatic states) may occur due to the inhibition of topoisomerase II activity during DNA synthesis [87, 88]. The local unraveling and subsequent rewinding of eu or heterochromatin regions undergoing replication require topoisomerase II activities to alleviate DNA torsional stress [86]. Etoposide stabilizes DNA-topoisomerase II cleavable complex and hinders the resealing of DSB introduced by the enzyme generating the accumulation of DSB unable to reach resolution. In addition, chromatin unwinding during replication may turn DNA more accessible to S-independent and S-dependent chemical agents including etoposide and MMS, respectively [91, 92].

3.2. Eu/heterochromatin replication and primary induced-damage distribution in interphase nuclei

Few minutes after exposure of mammalian cells to DSB-inducing agents, the nucleosomal histone variant H2AX is phosphorylated at serine 139 (humans) or 129 (mouse) of C-terminal tails reaching a peak of phosphorylation 30 min later. H2AX phosphorylation (named γH2AX) initiates around the induced DSB and spreads through a large chromatin region (~2000 H2AX molecules) flanking the lesion, which can be visualized as discrete γH2AX foci in interphase nuclei and mitotic chromosomes by means of specific fluorochrome-conjugated antibodies [99].

γH2AX is involved in the DDR by coordination with other damage response proteins to recruit signaling, remodeling, checkpoint and repair proteins. At sites of DSB, the DNA-PK (**DNA D**ependent **P**rotein **K**inase) binds to activate the **N**on **H**omologous **E**nd **J**oining (NHEJ) DSB repair pathway. If DSB are produced after replication, RAD51 and BRCA2 are recruited to DSB sites initiating the **H**omologous **R**ecombination repair pathway (HR). Simultaneously, the sensing complex MRN (MRE11, RAD50, NBS1) associates to DSB, facilitating the recruitment and activation (auto-phosphorylation) of ATM (**A**taxia **T**elangiectasia **M**utated), MDC1, BRCA1 and 53BP1 [100].

ATM, ATR (**ATM**- and **R**ad3-related) and DNA-PK are members of the phosphatidylinositol 3-kinase-like family of serine/threonine protein kinases that phosphorylate H2AX. Unlike ATM, which appears to be mainly activated by DSB, ATR seems to be activated by induced SSB and the excision repair SSB intermediates generated during DNA repair. Since ATR activation was observed in replicating cells, it was suggested that the blockage of replication forks by SSB is required to initiate ATR-mediated phosphorylation of H2AX. Besides, it was reported that stalled replication forks may also trigger H2AX phosphorylation when bulky lesions (i.e.: CPD and 6-4 PP) collide with replication forks [101, 102].

NBS1, MDC1, 53BP1, and BRCA1 may all function as mediators and amplifiers of the DDR, recruiting diverse repair and checkpoint proteins (including ATM and ATR) and generating an amplification loop that also extends H2AX phosphorylation [99]. 53BP1 can bind directly to H3K79me and H4K20me accumulated at sites of DSB collaborating with a global chromatin unwinding following the formation of DSB in concert with other proteins like TIP60 (member of an histone acetyltransferases family) and KAP1 [103, 104, 105].

Several immunofluorescence studies have demonstrated that induced-γH2AX foci are located preferentially within euchromatic regions of the genome, suggesting that heterochromatin could be refractory to γH2AX foci formation. Employing immuno-FISH to analyze radiation induced-DSB (γH2AX foci) in chromatin regions with known chromatin compaction (human chromosome 18 versus chromosome 19; RIDGE versus anti-RIDGE region of human chromosome 11), it has been observed that condensed regions of gene-poor chromatin are less susceptible to DSB induction compared with decondensed, gene-rich chromatin [106-109].

Different hypothesis have been raised to explain the non-homogeneous distribution of γH2AX foci in nuclei. The highly condensed state or abundance of binding proteins may reduce the accesibility of chemical DNA damaging agents to heterochromatin. Besides, since condensed chromatin is less hydrated than euchromatin, a lower amount of free radicals could be induced by radiation [110]. Furthermore, compact heterochromatin could contain a lower proportion of H2AX isoform or be less accessible to kinases due to compaction or protein coating [106]. Additionally, a wave of chromatin unwinding starting at DSB sites and spreading throughout the entire chromatin was described (as a result of KAP1 phosphorylation by ATM kinase) implying that the preferential location of γH2AX foci in decondensed chromatin perhaps reflects chromatin reorganization [105, 111-113].

Finally, a short-range migration of DSB from packed chromatin toward specific decondensed DSB repair domains could also take place [106, 110]. Using carbon ion microirradiation to induce DSB combined to a modified TUNEL assay to directly visualize these lesions and γH2AX immunodetection, a bending of the linear ion-induced γH2AX track around heterochromatic regions was observed [114]. The γH2AX foci migration from the interior to the periphery of heterochromatin appears to initiate within 20 min post-irradiation and be almost complete 1 h after damage induction. The decondensation of heterochromatin at sites of ion hits possibly promotes the movement of DSB to peripheral regions of lower chromatin density where repair may potentially proceed [114].

To assess the influence of replication in the distribution of chromatin damage, we analyzed the localization of bleomycin-induced γH2AX foci in relation to replication of eu- or heterochromatin interphase compartments in 5-ethynyl-2′deoxiuridine (EdU) pulsed-labeled CHO9 nuclei. Bleomycin (BLM) is a radiomimetic S-independent clastogen that induces oxidative damage, SSB and mainly DSB as well as a rapid phosphorylation of H2AX [115].

Asynchronously growing CHO9 cultures were pulse-exposed (30 min) to EdU (controls) or simultaneously (30 min) treated with BLM (40 μg/ml). Early and late replication regions and γH2AX foci were detected with an azide conjugated to Alexa Fluor 488 (Click-iT EdU, Invitrogen) and mouse anti-γH2AX (Abcam) followed by Cy3-conjugated antimouse antibodies, respectively. Single-cell z-stacks from control (n=25) and treated (n=63) nuclei were obtained by confocal microscopy and processed with Image J software. Using binary masks for each channel, the relation (ratio) between the percentage of damaged (γH2AX) area in replicating chromatin (EdU) area and the percentage of damaged area in the whole nuclear area (DAPI) was calculated for each nucleus. Finally, the arithmetic mean of the ratios corresponding to early S (n=30) and late S (n=33) nuclei was calculated.

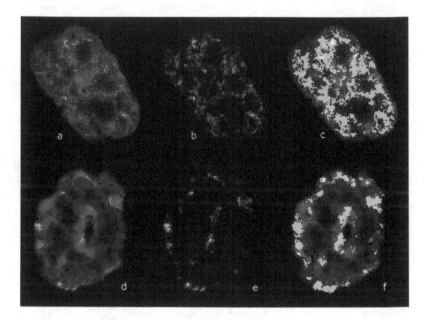

Figure 7. Distribution of BLM induced-γH2AX foci revealed by immunolabelling (Cy3; red) in early (top) or late (bottom) S-phase CHO9 nuclei. Replicating patterns were obtained by EdU incorporation and chemical detection (azide-Alexa Fluor 488; green). Nuclei were counterstained with DAPI (blue). Early S (a-c) and late S replicating nuclei are shown. Panels (a, d) and (d, e) contain DAPI/γH2AX/EdU and γH2AX/EdU merged images, respectively. Panels (c) and (f) illustrate binary masks of red (γH2AX) and green (EdU) channels overlaying the respective DAPI images.

Preliminary results (arithmetic mean of the ratios: 1.57 in early S- and 1.45 in late S-nuclei) suggest a bias in damage distribution towards replicating areas (~50 % higher than expected) probably due to local unwinding of chromatin down to naked DNA in both eu- and heterochromatin during DNA synthesis. Chromatin decondensation may increase the susceptibility to DNA damage as well as the accessibility of kinases that phosphorylate H2AX. Noteworthy, detailed visual analysis of fluorescent images or the corresponding binary masks in both early and late S-phase revealed that these results were not due to a large amount of γH2AX foci dwelling within replicating area and few of them outside. Instead, γH2AX foci recurrently mapped to the interfaces between replicating and non-replicating regions (Figure 7; Liddle P, unpublished observations).

The fact that in late-replicating cells γH2AX foci tend to map to the boundaries of replicating compartments (Figure 7, panels d-f) may be due to repositioning of damaged sites to less condensed peripheral heterochromatin regions, as it has been suggested in other models [112, 113]. However, this peculiar distribution of γH2AX foci in replicating/non-replicating interfaces was also observed in early S-phase when the less compact euchromatin replicates (Figure 7, panels a-c). In this respect, BLM-induced DNA lesions could preferentially map at the damage-prone TTR located at the boundaries of early and late replicating compartments.

4. Conclusions

We assayed the influence of eu/heterochromatin replication timing in the distribution of chromatin induced damage using two different approaches: (1) the analysis of UV-C, MMS and etoposide-induced BP in Xp or Xq replicating CHO9 X mitotic chromosome and; (2) the analysis of primary BLM-induced damage (γH2AX foci) in CHO9 early and late replicating interphase nuclei. Our findings support the assumption that induced damage patterns shift according to eu- or heterochromatin replication. The asynchronic replication of eu- or heterochromatin compartments could influence the distribution of primary DNA lesions and CA, prevailing in replicating chromatin regions, irrespective of its eu- or heterochromatic state. Thus, eu/heterochromatin replication timing seems to play an overriding role in the production and localization of chromosome damage in S-phase cells.

Acknowledgments

We are indebted to the PEDECIBA Postgraduate Program, the National Agency of Investigation and Innovation (ANII) and the Alexander von Humboldt Foundation (AvH). Liddle P. is a former Fellow of the AvH Förderung Program at the LMU Biozentrum (Munich).

Author details

María Vittoria Di Tomaso, Pablo Liddle, Laura Lafon-Hughes, Ana Laura Reyes-Ábalos and Gustavo Folle

*Address all correspondence to: marvi@iibce.edu.uy

Department of Genetics, Instituto de Investigaciones Biológicas Clemente Estable, Montevideo, Uruguay

References

[1] Gilbert DM. Replication timing and transcriptional control: beyond cause and effect. Current Opinion in Cell Biology 2002;14 377-383.

[2] Gilbert DM, Gasser SM. Nuclear structure and DNA replication. In: DePamphilis ML. (ed.) DNA replication and human disease. New York: Cold Spring Harbor Laboratory Press; 2006. p175-196.

[3] Heitz E. Das Heterochromatin der Moose. Jahrbuch der Wissenschafltichen Botanik 1928;69 762-818

[4] Brown SW. Heterochromatin. Science 1966;151 417-425.

[5] Craig JM. Heterochromatin-many flavours, common themes. BioEssays 2004;27 17-28.

[6] Holmquist GP, Ashley T. Chromosome organization and chromatin modification: influence on genome function and evolution (Review). Cytogenetic and Genome Research 2006;114(2) 96-125.

[7] Gilbert N, Boyle S, Fiegler H, Woodfine K, Carter NP, Bickmore WA. Chromatin architecture of the human genome: gene-rich domains are enriched in open chromatin fibers. Cell 2004;118 555-566.

[8] Arrighi FE, Hsu TC. Localization of heterochromatin in human chromosomes. Cytogenetics and Cell Genetics 1971;10 81-86.

[9] Puck TT, Cieciura SJ, Robinson A. Genetics of somatic mammalian cells. III. Long term cultivation of euploid cells from human and animal subjects. The Journal of Experimental Medicine 1958;108 954-956.

[10] Drets ME, Shaw MW. Specific banding patterns of human chromosomes. Proceedings of the National Academy of Sciences of the United States of America 1971;68 2073-2077.

[11] Holmquist GP. DNA sequences in G-bands and R-bands. In: Adolph K W (ed.) Chromosomes and Chromatin. Boca Raton: CRC Press; 1988; p76-121.

[12] Saitoh Y, Laemmli UK. Metaphase chromosome structure: bands arise from a differential folding path of the highly AT-rich scaffold. Cell 1994;76 609-622.

[13] Korenberg JR, Rykowski MC. Human genome organization: Alu, Lines, and the molecular structure of metaphase chromosome bands. Cell 1988;53 391-400.

[14] Yasuhara JC, Wakimoto BT. Oxymoron no more: the expanding world of heterochromatic genes. Trends in Genetics 2006;22(6) 330-338.

[15] Dimitri P, Caizzi R, Giordano E, Accardo MC, Lattanzi G, Biamonti G. Constitutive heterochromatin: a surprising variety of expressed sequences. Chromosoma 2009;118 419-435.

[16] Sadoni N, Langer S, Fauth C, Bernardi G, Cremer T, Turner BM, Zink D. Nuclear organization of mammalian genomes: polar chromosome territories build up functionally distinct higher order compartments. The Journal of Cell Biology 1999;146 1211-1226.

[17] Jenuwein T, Allis CD. Translating the histone code. Science 2001;293 1074-1080.

[18] Pokholok DK, Harbison CT, Levine S, Cole M, Hannett NM, Lee TI, Bell GW, Walker K, Rolfe PA, Herbolsheime E. Genome-wide map of nucleosome acetylation and methylation in yeast. Cell 2005;122 517-527.

[19] Kim JH, Workman JL. Histone acetylation in heterochromatin assembly. Genes and Development 2010;24 738-740.

[20] Ohlsson R, Renkawitz R, Lobanenkov V. CTCF is a uniquely versatile transcription regulator linked to epigenetics and disease. Trends in Genetics 2001; 17 520-527.

[21] Woodfine K, Fiegler H, Beare DM, Collins JE, McCann OT, Young BD, Debernardi S, Mott R, Dunham I, Carter NP. Replication timing of the human genome. Human Molecular Genetics 2004;13 191-202.

[22] Gilbert DM. Evaluating genome-scale approaches to eukaryotic DNA replication. Nature Reviews Genetics 2010;11 673-684.

[23] Ryba T, Hiratani I, Lu J, Itoh M, Kulik M, Zhang J, Dalton S, Gilbert D M. Evolutionarily conserved replication timing profiles predict long-range chromatin interactions and distinguish closely related cell types. Genome Research 2010;20 761-770.

[24] Farkash-Amar S, Lipson SD, Polten A, Goren A, Helmstetter C, Yakhini Z,. Simon I. Global organization of replication time zones of the mouse genome. Genome Research 2008;18 1562-1570.

[25] Maya-Mendoza A, Petermann E, Gillespie DA, Caldecott KW, Jackson DA. Chk1 regulates the density of active replication origins during the vertebrate S phase. The EMBO Journal 2007;26 2719-2731.

[26] Labit H, Perewoska I, Germe T, Hyrien O, Marheineke K. DNA replication timing is deterministic at the level of chromosomal domains but stochastic at the level of replicons in Xenopus egg extracts. Nucleic Acids Research 2008;36 5623-5634.

[27] Ma H, Samarabandu J, Devdhar RS, Acharya R, Cheng PC, Meng C, Berezney R. Spatial and temporal dynamics of DNA replication sites in mammalian cells. The Journal of Cell Biology 1998;143 1415-1425.

[28] Jeon Y, Bekiranov S, Karnani N, Kapranov P, Ghosh S, MacAlpine D, Lee C, Hwang DS, Gingeras TR, Dutta A. Temporal profile of replication of human chromosomes. Proceedings of the National Academy of Sciences of the United States of America 2005;102 6419-6424.

[29] Hiratani I, Leskovar A, Gilbert DM. Differentiation-induced replication-timing changes are restricted to AT-rich/long interspersed nuclear element (LINE)-rich iso-

chores. Proceedings of the National Academy of Sciences of the United States of America 2004;101(48) 16861-16866.

[30] Subramanian PS, Chinault AC. Replication timing properties of the human HPRT locus on active, inactive and reactivated X chromosomes. Somatic Cell and Molecular Genetics 1997;23 97-109.

[31] Kim SM, Dubey DM, Huberman JA. Early-replicating heterochromatin. Genes and Development 2003;17 330-335.

[32] Wright WE, Tesmer VM, Liao ML, Shay JW. Normal human telomeres are not late replicating. Experimental Cell Research 1999;251 492-499.

[33] Holló G, Keresõ J, Praznovszky T, Cserpán I, Fodor K, Katona R, Csonka E, Fatyol K, Szeles A, Szalay AA, Hadlaczky G. Evidence for a megareplicon covering megabases of centromeric chromosome segments. Chromosome Research 1996;4 240-247.

[34] Schübeler D, Scalzo D, Kooperberg C, van Steensel B, Delrow J, Groudine MGenome-wide DNA replication profile for Drosophila melanogaster: a link between transcription and replication timing. Nature Genetics 2002;32 438-442.

[35] Wutz A, Jaenisch R. A shift from reversible to irreversibleX inactivation is triggered during ES cell differentiation. Molecular Cell 2000;5 695-705.

[36] Dimitrova DS, Gilbert DM. The spatial position and replication timing of chromosomal domains are both established in early G1-phase. Molecular Cell 1999;4 983-993.

[37] Arney KL, Fisher AG. Epigenetic aspects of differentiation. Journal of Cell Science 2004;117 4355-4363.

[38] Zhou J, Ermakova OV, Riblet R, Birshtein BK, Schildkraut CL. Replication and sub-nuclear location dynamics of the immunoglobulin heavy-chain locus in B-lineage cells. Molecular Cell Biology 2002a;22 4876-4889.

[39] Taddei A, Van Houwe G, Hediger F, Kalck V, Cubizolles F, Schober H, Gasser SM. Nuclear pore association confers optimal expression levels for an inducible yeast gene. Nature 2006; 441 774-778.

[40] Mendjan S, Taipale M, Kind J, Holz H, Gebhardt P, Schelder M, Vermeulen M, Buscaino A, Duncan, K, Mueller J, Wilm M, Stunnenberg HG, Saumweber H, Akhtar A. Nuclear pore components are involved in the transcriptional regulation of dosage compensation in Drosophila. Molecular Cell 2006;21 811-823.

[41] Goldman MA, Holmquist GP, Gray MC, Caston LA, Nag A. Replication timing of mammalian genes and middle repetitive sequences. Science 1984;224 686-692.

[42] Craig JM, BickmoreWA. Chromosome bands-flavours to savour. BioEssays 1993;15 349-354.

[43] Takebayashi S, K. Sugimura, T. Saito, C. Sato, Y. Fukushima, H. Taguchi, and Okumura K. Regulation of replication at the R/G chromosomal band boundary and peri-

centromeric heterochromatin of mammalian cells. Experimental Cell Research 2005;304 162-174.

[44] Méndez J. Temporal regulation of DNA replication in mammalian cells. Critical Reviews in Biochemistry and Molecular Biology 2009;44 343-351.

[45] McNairn AJ, Gilbert DM. Epigenomic replication: linking epigenetics to DNA replication. BioEssays 2003;25 647-656.

[46] Casas-Delucchi CS, van Bemmel JG, Haase S, Herce HD, Nowak D, Meilinger D, Stear JH, Leonhardt H, Cardoso MC. Histone hypoacetylation is required to maintain late replication timing of constitutive heterochromatin. Nucleic Acids Research 2011;10 1-11.

[47] Unnikrishnan A,. Gafken PR, Tsukiyama T. Dynamic changes in histone acetylation regulate origins of DNA replication. Nature Structural and Molecular Biology 2010;17 430–437.

[48] Bickmore WA, Carothers AD. Factors affecting the timing and imprinting of replication on a mammalian chromosome. Journal of Cell Science 1995;108 2801-2809.

[49] Craig JM, Wong LH, Lo AWI, Earle E, Choo KHA. Centromeric chromatin pliability and memory at a human neocentromere. The EMBO Journal 2003;12 3109-3121.

[50] Cowell IG, Aucott R, Mahadevaiah Sk, Burgoyne PS, Huskisson N, Bongorini S, Prantera G, Fanti L, Pimpinelli S, Wu R, Gilbert DM, Shi W, Fundele R, Morrison H, Jeppesen P, Singh PB. Heterochromatin, HP1 and methylation at lysine 9 of histone H3 in animals. Chromosoma 2002;111 22-36.

[51] Hediger F, Gasser SM. Heterochromatin protein 1: don't judge the book by its cover! Current Opinion in Cell Biology 2006;16 143-150.

[52] Hansen RS, Stoger R, Wijmenga C, Stanek AM, Canfield TK, et al. 2000 Escape from gene silencing in ICF syndrome: evidence for advanced replication time as a major determinant. Human Molecular Genetics 2000;9 2575-2587.

[53] Rhind N, Yang SC, Bechhoefer J. Reconciling stochastic origin firing with defined replication timing. Chromosome Research 2010;18 35-43.

[54] Yang, S. C., N. Rhind, and J. Bechhoefer. Modeling genome-wide replication kinetics reveals a mechanism for regulation of replication timing. Molecular Systems Biology 2010;6:404.

[55] Smith ZE, Higgs DR. The pattern of replication at a human telomeric region (16p13.3): its relationship to chromosome structure and gene expression. Human Molecular Genetics 1999;8 1373-86.

[56] Costanzo V, Shechter D, Lupardus PJ, Cimprich KA, Gottesman M, Gautier J. An ATR- and Cdc7-dependent DNA damage checkpoint that inhibits initiation of DNA replication. Molelular Cell 2003;11 203-213.

[57] Herrick J. Genetic variation and DNA replication timing, or why is there late replicating DNA? Evolution 2011;65 3031-3047.

[58] Niida H, Katsuno Y, Banerjee B, Hande MP, Nakanishi M. Specific role of Chk1 phosphorylations in cell survival and checkpoint activation. Molecular Cell Biology 2007; 27 2572-2581.

[59] Karnani N, Dutta A. The effect of the intra-S-phase checkpoint on origins of replication in human cells. Genes and Development 2011;25 621-633.

[60] Machida YJ, Hamlin JL, Dutta A. Right place, right time, and only once: replication initiation in metazoans. Cell 2005;123 13-24.

[61] Loupart ML, Krause SA, Heck MMS. Aberrant replication timing induces defective chromosome condensation in Drosophila ORC2 mutants. Current Biology 2000;10 1547-1556.

[62] Pflumm MF. The role of DNA replication in chromosome condensation. BioEssays 2002;24 411-418.

[63] Sasaki T, Gilbert DM. The many faces of the origin recognition complex. Current Opinion in Cell Biology 2007;19 337-343.

[64] Gardner RJM, Sutherland GR. Chromosome Abnormalities and Genetic Counseling. New York: Oxford University Press; 1966.

[65] Mitelman F. Patterns of chromosome variation in neoplasia. In: Obe G, Natarajan AT (ed.) Chromosomal Aberrations: Basic and Applied Aspects. Berlin: Springer-Verlag; 1990; p 86-100.

[66] De Braekeleer M. Fragile sites and chromosomal structural rearrangements in human leukemia and cancer. Anticancer Research 1987;7 417-422.

[67] Yunis JJ, Soreng AL, Bowe AE. Fragile sites are target of diverse mutagens and carcinogens. Oncogene 1987;1 59-69.

[68] Hecht F. Fragile sites, cancer chromosome breakpoints and oncogenes all cluster in light G bands. Cancer Genetics and Cytogenetics 1988; 31 17-24.

[69] Barrios L, Miró R, Caballín MR, Fuster C, Guedea F, Subias A, Egozcue J. Cytogenetic. effects of radiotherapy breakpoint distribution in induced chromosome aberrations. Cancer Genetics and Cytogenetics 1989;41 61-70.

[70] Porfirio B, Tedeschi B, Vernole P, Caporossi D, Nicoletti B The distribution of Msp I-induced breaks in human lymphocyte chromosomes and its relationship to common fragile sites. Mutation Research 1989;213 117-124.

[71] Tedeschi B, Porfirio B, Caporossi D, Vernole P, Nicoletti B. Structural chromosomal rearrangements in Hpa II-treated human lymphocytes. Mutation Research 1991;248 115-121.

[72] Folle G, Liddle P, Lafon-Hughes L, Di Tomaso M. Close encounters: RIDGEs, hyper-acetylated chromatin, radiation breakpoints and genes differentially expressed in tumors cluster at specific human chromosome regions. Cytogenetics and Genome Research 2010; 128 17-27.

[73] Gollin SM. Mechanisms leading to chromosomal instability. Seminars in Cancer Biology 2005;15 33-42.

[74] Folle GA. Nuclear architecture, chromosome domains and genetic damage. Mutation Research-Reviews in Mutation Research 2008;658 172-183.

[75] Ermakova OV, Nguyen LH, Little RD, Chevillard C, Riblet R, Ashouian N, Birshtein BK, Schildkraut CL. Evidence that a single replication fork proceeds from early to late replicating domains in the IgH locus in a Non-B cell line. Molecular Cell 1999;3 321-330.

[76] Durkin S, Glover T. Chromosome fragile sites. Annual Review of Genetics 2007;41 169-192.

[77] Letessier A, Millot GA, Koundrioukoff S, Lachagès AM, Vogt N, Hansen RS, Malfoy B, Brison O, Debatisse M. Cell-types pecific replication initiation programs set fragility of the FRA3B fragile site. Nature 2011;470 120-123.

[78] Flynn KM, Vohr SH, Hatcher PJ, Cooper VS. Evolutionary rates and gene dispensability associate with replication timing in the archaeon Sulfolobus islandicus. Genome Biology and Evolution 2010;2 859-869.

[79] Mugal CF,. Wolf JB, von Grünberg HH, Ellegren H. Conservation of neutral substitution rate and substitutional asymmetries in mammalian genes. Genome Biology and Evolution 2010;2 19-28.

[80] Kaufman DG, Cohen SM, Chastain PD. Temporal and functional analysis of DNA replicated in early S phase. Advances in Enzyme Regulation 2011;51 257-271.

[81] Elango N, Kim SH, Vigoda E, Yi SV. Mutations of different molecular origins exhibit contrasting patterns of regional substitution rate variation. PLoS Computational Biology 2008;4(2) e1000015. doi:10.1371/journal.pcbi.1000015

[82] Walser JC, Furano AV. The mutational spectrum of non-CpG DNA varies with CpG content. Genome Research 2010;20 875-882

[83] Nyberg KA, Michelson RJ, Putnam CW, Weinert TA. DNA damage and replication checkpoints. Annual Review of Genetics 2002;36 617-656

[84] Di Tomaso MV, Martínez-López W, Méndez- Acuña L, Lafon-Hughes L, Folle GA. Factors leading to the induction and conversion of DNA damage into structural chromosomal aberrations. In: Miura S, Nakano S (ed.). Progress in DNA Damage Research. New York: Nova Publisher; 2008; p30-40.

[85] Obe G, Johannes C, Schulte-Frohlinde D. DNA double-strand breaks induced by sparsely ionizing radiation and endonucleases as critical lesions for cell death, chro-

mosomal aberrations, mutations and oncogenic transformation. Mutagenesis 1992;7 3-12.

[86] Champoux JJ. DNA topoisomerases: structure, function, and mechanism. Annual Review of Biochemistry 2001;70 369-413.

[87] Palitti F, Mosesso P, Di Chiara D, Schinoppi A, Fiore M, Bassi L. Use of antitopoisomerase drugs to study the mechanisms of induction of chromosomal damage. In: Chromosomal Alterations: Origin and Significance. Obe G, Natarajan AT (ed.) Berlin: Springer-Verlag; 1994; p103-115.

[88] Degrassi F, Fiore M, Palitti F. Chromosomal aberrations and genomic instability induced by topoisomerase-targeted antitumor drugs. Current Medicinal Chemistry-Anti-Cancer Agents 2004;4 317-325.

[89] Ray M, Mohandas T. Proposed banding nomenclature for the Chinese hamster chromosomes (Cricetulus griseus). In: Report of the Committee on Chromosome Markers. Hamerton JL (ed.) Cytogenetics and Cell Genetics 1976;16 83-91.

[90] Schmid W, Leppert MF. Rates of DNA synthesis in heterochromatic and euchromatic segments of the chromosome complements of two rodents. Cytogenetics 1969;8 125-135.

[91] Di Tomaso MV, Martínez-López W, Folle GA, Palitti F. Modulation of chromosome damage localisation by DNA replication timing. International Journal of Radiation Biology 2006;82 877-886.

[92] Di Tomaso MV, Martínez-López W, Palitti F. Asynchronously replicating eu/heterochromatic regions shape chromosome damage Cytogenetics and Genome Research 2010;1281(1-3) 111-117.

[93] De Boer J, Hoeijmakers JH. Nucleotide excision repair and human syndromes. Carcinogenesis 2000;21 453-460.

[94] García CL, Carloni M, de la Pena NP, Fonti E, Palitti F. Detection of DNA primary damage by premature chromosome condensation in human peripheral blood lymphocytes treated with methylmethanesulfonate. Mutagenesis 2001;16 121-125.

[95] Christmann M, Roos WP, Kaina B. DNA methylation damage formation, repair and biological consequences. In Obe G, Vijayalaxmi, (ed.) Chromosomal Alterations: Methods, Results and Importance in Human Health. Heidelberg: Springer Verlag; 2007; p99-121.

[96] Bender MA, Griggs HG, Walker PL. Mechanisms of chromosomal aberration production. I. Aberration induction by ultraviolet light. Mutation Research 1973;20 387-402.

[97] Natarajan AT, Obe G, van Zeeland AA, Palitti F, Meijers M, Vergaal-Immerzeel EAM. Molecular mechanisms involved in the production of chromosomal aberrations. II. Utilization of [96]Neurospora endonuclease for the study of aberration pro-

duction by X-rays in G1 and G2 stages of the cell cycle. Mutation Research 1980;69 293- 305.

[98] Bradley M. Double-strand breaks in DNA caused by repair of damage due to ultra-violet light. Journal of Supramolecular Structure and Cellular Biochemistry 1981;(16) 337-343.

[99] Rogakou EP, Boon C, Redon C, Bonner WM. Megabase chromatin domains involved in DNA double-strand breaks in vivo. The Journal of Cell Biology 1999;146(5) 905-916.

[100] Li L, Zou L. Sensing, signaling, and responding to DNA damage: organization of the checkpoint pathways in mammalian cells. Journal of Cellular Biochemistry 2005;4(2) 298-306.

[101] Marti TM, Hefner E, Feeney L, Natale V, Cleaver JE. H2AX phosphorylation within the G1phase after UV irradiation depends on nucleotide excision repair and not DNA double-strand breaks. Proceedings of the National Academy of Sciences of the United States of America 2006;103 9891-9896.

[102] Hanasoge S, Ljungman M. H2AX phosphorylation after UV irradiation is triggered by DNA repair intermediates and is mediated by the ATR kinase. Carcinogenesis 2007;28 2298-2304.

[103] Huyen Y, Zgheib O, Ditullio RA Jr, Gorgoulis VG, Zacharatos P, Petty TJ. Methylated lysine 79 of histone H3 targets 53BP1 to DNA double-strand breaks. Nature 2004;432 406-411.

[104] Botuyan MV, Lee J, Ward IM, Kim JE, Thompson JR, Chen J, Mer G. Structural basis for the methylation state-specific recognition of histone H4-K20 by 53BP1 and Crb2 in DNA repair. Cell 2006;7 1361-1373.

[105] Noon AT, Shibata A, Rief N, Löbrich M, Stewart GS, Jeggo PA, Goodarzi AA. 53BP1-dependent robust localized KAP-1 phosphorylation is essential for heterochromatic DNA double-strand break repair. Nature Cell Biology 2010;12 177-184.

[106] Cowell IG, Sunter NJ, Singh PB, Austin CA, Durkacz BW, Tilby MJ. GammaH2AX foci form preferentially in euchromatin after ionising-radiation. PLoS One 2007;2 e1057.

[107] Kim JA, Kruhlak M, Dotiwala F, Nussenzweig A, Haber J E. Heterochromatin is re-fractory to γ-H2AX modification in yeast and mammals. The Journal of Cell Biology 2007;178(29) 209-218.

[108] Falk M, Lukásová E, Kozubek S. Chromatin structure influences the sensititivty of DNA to γ-radiation. Biochimica et Biophysica Acta 2008;1783 2398-2414.

[109] Vasireddy RS, Karagiannis TC, Assam EO. -radiation-induced ⊛H2AX formation oc-curs preferentially in actively transcribing euchromatic loci. Cellular and Molecular Life Sciences 2010;67 291-294.

[110] Falk M, Lukásová E, Kozubek S. Higher-order chromatin structure in DSB induction, repair and misrepair. Mutation Research 2010;704 88-100.

[111] Ziv Y, Bielopolski D, Galanty Y, Lukas C, Taya Y, Schultz DC, Lukas J, Bekker-Jensen S, Bartek J, ShilohY. Chromatin relaxation in response to DNA double-strand breaks is modulated by a novel ATM- and KAP-1 dependent pathway. Nature Cell Biology 2006;8 870-876.

[112] Kruhlak MJ, Celeste A, Dellaire G, Fernandez-Capetillo O, Muller WG, et al. (2006) Changes in chromatin structure and mobility in living cells at sites of DNA double-strand breaks. The Journal of Cell Biology 2006;172 823-834.

[113] Cann K L, Dellaire G. Heterochromatin and the DNA damage response: the need to relax1. Biochemistry and Cell Biology 2011;89 45-60.

[114] Jakob B, Splinter J, Conrad S, Voss KO, Zink D, Durante M, Löbrich M, Taucher-Scholz1 G. DNA double-strand breaks in heterochromatin elicit fast repair protein recruitment, histone H2AX phosphorylation and relocation to euchromatin. Nucleic Acids Research 2011;39(15) 6489-6499.

[115] Chen J, Stubbe J Bleomycins: Toward better therapeutics. Nature Reviews Cancer 005;5 102-112.

Chapter 6

A Histone Cycle

Douglas Maya, Macarena Morillo-Huesca,
Lidia Delgado Ramos, Sebastián Chávez and
Mari-Cruz Muñoz-Centeno

Additional information is available at the end of the chapter

1. Introduction

Each time a cell divides it must duplicate its DNA content and segregate it equally between two daughter cells. Once a cell has decided to replicate its DNA, hundreds of different proteins must carefully interact with each other in a very orchestrated process. Defects in any of these steps can lead to cell death and genetic instability and have been shown to be present in many human diseases including cancer [1]. Each step of DNA replication must take place in a correct spatial and temporal window, so cells have evolved complex regulatory networks allowing an efficient regulation of this process. One important feature of eukaryotic cells is that DNA is strongly associated with histones, basic proteins that wrap DNA around octameric structures called nucleosomes. The association of DNA and nucleosomes is commonly known as chromatin.

Nucleosomes are, among others, one of the principal differences between eukaryotic and prokaryotic DNA. Unlike bacteria, eukaryotic cells are not able to live without DNA packed into chromatin [2]. Replication involves dramatic changes in the whole chromatin landscape, since nucleosomes must be removed transiently from the front of the replication machinery and repositioned after it. Nucleosome disassembly and assembly involves the action of chromatin remodeling factors, proteins able to destabilize interactions between histones and DNA allowing the interaction of other complexes with DNA. Restoration of chromatin after the replication bubble is a very important step because nucleosomes are not repetitive units of information and contain a specific epigenetic signature or code [3]. In order to ensure enough histones for the nascent DNA, cells must increase the pool of free histones. In human cells, each passage through S-phase requires the synthesis and assembly of almost 30 million nucleosomes that are synthesized mainly during S-phase and are rapidly packaged to DNA. Histone production is very tightly coupled to DNA synthesis and is rapidly shut off when replication finishes or is halted by treatment with mutagenic agents. Histone regulation is very important and accumulation of free histones in the cell has been shown to be highly deleterious for the cell and lead to chromosome loss [2].

Chromatin is a structural barrier for replication but can also play an important role in the regulation of some of the steps within. In this chapter, we will focus on how the chromatin landscape influences DNA replication and show that histones and DNA must adapt to each other in order to ensure a correct genomic duplication. We will describe the influence that chromatin plays at the different stages of DNA replication and then jump to the accurate control that cells exert on histone levels during the cell cycle. We will finally show different situations that uncouple DNA replication from histone deposition and synthesis, and discuss if chromatin state can influence the decision of cells to replicate their DNA or not.

2. Replication from a chromatin point of view

2.1. Chromatin influences early steps of replication

The initiation of DNA replication in any organism requires a series of proteins able to recruit and ultimately load two hexameric DNA helicases. These proteins are able to unwind DNA, a process required to start replication. In eukaryotic cells the pre-initiation complex is formed by two MCM2-7 rings that are loaded in an inactive form next to the Origin Recognition Complex (ORC). The MCM helicase must be activated by the sequential action of Dbf4 Dependent Kinase (DDK) and Cyclin Dependent Kinase (CDK), and the addition of other accessory proteins. In mammalian cells, 30,000 to 50,000 origins are sequentially activated each time a cell divides [4].

The nature of the sequence and the structure of replication origins is still a matter of debate, and most higher eukaryotes lack a specific consensus sequence for ORC binding. Origins of DNA Replication Initiation (ORIs) are normally regions of DNA sequence rich in AT that contain a nucleosome-free region (NFR) [5] [6] and it has been suggested that the chromatin environment is important for the establishment of the ORC complex [7] [8]. In *Drosophila* follicle cells histones that localize around ORIs are hyperacetylated and changes in the acetylation levels of these histones affects ORC binding [9].

As ilustrated in figure 1, methylation of histone H4 has also been shown to be important for ORC recruitment and artificial tethering of the methyltransferase SET8 to a random locus promotes ORC1 binding [10]. Once ORC is bound to DNA, proteins CDC6 and CDT1 help to load the two MCM2-7 helicases to DNA [11]. Loading could also be influenced by the acetylation of histone H4, since CDT1 is able to recruit the histone acetyltransferase HBO1 to the ORC and enhances the recruitment of MCM2-7 to the origin [12, 13] [14]. Nevertheless, some ORC and MCM subunits are acetylated by Hbo1 in yeast and could therefore be the real targets of this enzyme [14]. Although all origins that are selected are able to load the pre-initiation complex, only one out of every ten will fire. Regulation of firing depends on the activation of the MCM helicases by sequential phosphorylation of some of its subunits by DDK and CDK kinases, that allow the recruitment of CDC45 and the GINS complex [15]. Once these proteins are loaded, the rest of the replication machinery is recruited and replication starts.

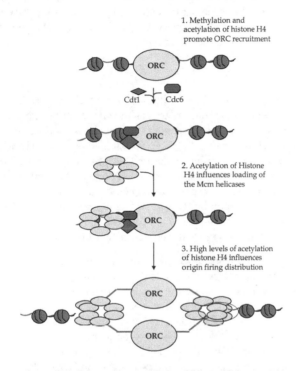

Figure 1. Chromatin influences DNA replication fork establishment. Schematic representation of the different steps during the assembly and activation of the replication fork machinery that are influenced by histone modifications. 1.ORC recruitment is influenced by methylation and acetylation levels of histone I-I4. 2.Acetylation of histone H4 by the Hbo1 might influence loading of the two Mcm2-7 helicases. 3. H4K 16 aoetytation is related to the timing of origin firing distribution.

Two interesting observations have lead to the hypothesis that chromosome architecture can also be important for origin usage. One of them is that origins seem to be organized into clusters of 5 to 10 origins that fire simultaneously [18]. Cohesins are enriched next to origins and depletion of the RAD21 cohesin subunit greatly reduces the number of active origins. This ring-like complex is able to wrap two chromatin fibers and creates a chromatin loop. It has been suggested that this spatial organization of chromatin could define replication domains that are activated synchronously [16, 17]. In agreement with this hypothesis, analysis of the oriGNA13 replication origin of hamster cells shows that active origins localize close to the base of chromatin loops [18]. The second feature that directly relates chromatin structure and replication is that genome replication does not take place in a single and continuous way. Domains containing several megabases of contiguous DNA are replicated earlier than others [19] and this replication timing is somehow correlated with acetylation levels of histone H4 at the K16 residue [20].

Chromatin influences the recruitment and the activity of different elements of the replication machinery. Once this machinery has been set up and is fully active, fork progression must now cope with the fact that approximately every 192 bp there is a nucleosome that must be displaced from DNA in order to continue with replication.

2.2. Nucleosome reorganization around the replication fork

Replication fork progression involves many proteins that interact closely with DNA. Electron micrography of replicating SV40 mini-chromosomes has shown that 300bp ahead of the replication fork, DNA remains *naked*, or at least contains nucleosomes that are unstable when compared to a canonical nucleosome [21] [22]. MCM progression in mammalian cells suggests that chromatin is decondensed in front of the replication fork [23] and artificial tethering of Cdc45 to DNA is able to promote a partial decondensation of chromatin without DNA synthesis [24]. This initial decondensation could be related to an increase in the mobility of histone H1 due to its phosphorylation by the cyclin A-CDK2 complex. It is still unclear if nucleosome disruption in front of the replication fork is due to specific chromatin remodeling in front of the fork or to the passage of the replication machinery itself [3].

Nucleosome disassembly and reassembly are processes quite well described for transcription. The efficiency of these processes is largely dependent on chromatin-remodeling complexes, proteins able to interact with and change the stability of chromatin, allowing the transcription machinery to interact with DNA. There are many different chromatin-remodeling complexes and all of them are possible candidates for nucleosome eviction during replication. The fact that chromatin disassembly and assembly occur in such a small spatial window makes it very difficult to distinguish between the complexes required for one or the other process. There are two major complexes that could be involved in H2A/H2B displacement during replication: FACT and NAP1, and another two for H3/H4: Asf1 and CAF1.

The FACT complex is composed of two main subunits, SPT16 and SSRP1 (Pob3 in *S.cerevisiae*), and plays a key role in nucleosome reorganization during transcription elongation. FACT function has been mostly related to the displacement of an H2A/H2B dimer during transcription, but it has also been proposed that displacement could be an indirect effect of nucleosome reorganization by this complex [25, 26]. There are many different items of evidence suggesting that FACT plays a role as a histone chaperone during DNA replication. FACT is required for DNA replication in *Xenopus* extracts, is present at human replication origins [27] and has been co-purified as part of the replication fork progression complex in yeast [28]. The other H2A/H2B histone chaperone candidate is Nap1, which has been shown to disassemble nucleosomes in vitro when combined with the RSC complex [29, 30]. Once H2A/H2B dimers are displaced, the H3/H4 tetramer is now accessible for an H3/H4 chaperone.

To date, it is not known if H2A/H2B dimers removed during replication are recycled. On the contrary, it is well established that the original H3/H4 tetramer present in front of the replication fork machinery is restored after the replication fork in a random semi-conservative process [31]. The fact that the H3/H4 tetramer is recycled suggests that a histone chaperone must disassemble this tetramer in front of the replication machinery and reassemble it after. One good candidate for this process is Asf1. This protein along with Chromatin Assembly Factor 1 (CAF1) plays a key role in deposition of new H3/H4 following passage of the replication fork. Asf1 binds PCNA and replication factor C [32], and can also bind the MCM helicase complex through histones H3 and H4. Upon a replication fork progression block, Asf1 can be found associated with post-translationally modified H3/H4 histones, which most likely belong to the parental chromatin [33].

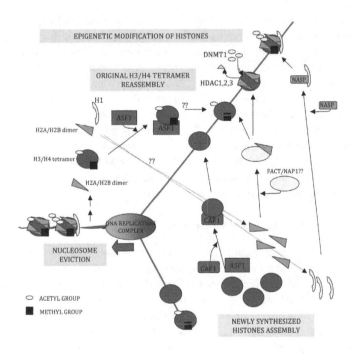

Figure 2. Nucleosome reorganization around the replication fork. Representation of the different nucleosome reorganization events that take place during replication fork progression. In order to simplify the figure, the DNA replication machinery and other accessory proteins that are important during fork progression are not shown. Interrogation marks are used when the protein/s involved in such process remain unknown or when the pathway has not been directly demonstrated.

Deposition of histone octamers occurs as soon as DNA is long enough to wrap around a nucleosome [21]. Since one H3/H4 tetramer is recycled after DNA replication, one new H3/H4 tetramer must be deposited on the other strand. The mechanism of reposition of the original

H3/H4 tetramer remains unclear but probably involves Asf1 (see previous paragraph). Incorporation of the new H3/H4 tetramer on the other hand is more defined and involves the action of CAF1 and Asf1. CAF1 is recruited to both leading and lagging strands by the proliferating cell nuclear antigen PCNA. Depletion of CAF1 produces a clear decrease in the assembly of new chromatin during replication [34], activates the DNA Damage Checkpoint (DDR), and stalls replication forks [35], suggesting that efficient chromatin repositioning after replication is important for replication fork progression. Asf1 plays a role in this process as a histone pool protein that delivers H3/H4 dimers to CAF1. After the H3/H4 tetramer is assembled, H2A/H2B becomes incorporated into chromatin in a process that probably involves FACT or NAP1. Finally, H1 protein is incorporated to allow further compacting of chromatin. Incorporation of H1 is probably mediated by the NASP protein and is required for efficient S-phase progression [36].

One interesting feature recently described is that the chromatin landscape influences the length of the Okazaki fragments synthesized at the lagging strand during DNA replication [37]. Due to the 5′ to 3′ polarity of DNA polymerase, synthesis of DNA in the lagging strand is discontinuous and generates short fragments of DNA named Okazaki fragments. These fragments must then join to form a unique DNA strand in a process known as maturation. Okazaki fragment maturation requires the sequential action of the flap endonuclease 1 (FEN1) and DNA ligase I. This group has demonstrated that the ligation junctions of Okazaki fragments are preferentially located in the nucleosome midpoint. The length of Okazaki fragments depends on the chromatin behind the replication fork and mutations that impair chromatin repositioning increase the average size of Okazaki fragments. According to their model, Pol · runs into the nucleosome assembled into the previous Okazaki fragment and this triggers termination, flap processing and ligation.

2.3. Chromatin maturation and centromere formation

After nucleosome incorporation to DNA, two major processes must take place: chromatin maturation and centromere formation. Histones start to acquire certain modifications in their tails as soon as they are repositioned to DNA. Nascent chromatin is highly acetylated and must be deacetylated and methylated to reach a more compact state. Deacetylation normally takes place by the histone deacetylases HDAC1-3 and DNA methylation by the DNA methyltransferase 1 (DNMT1). In addition to chromatin compacting, there are also some specific post-transcriptional modifications that must be acquired to establish a specific epigenetic code that is transmitted to daughter cells. Maintaining this "epigenetic memory" of daughter cells is important and has implications during cell differentiation (23). Restoration of all these marks does not take place exclusively in replication and can also take place during mitosis or even in daughter cells [38, 39]. Replication of the chromatin near the centromere is also vital to ensure an efficient segregation during mitosis. This heterochromatin presents a specific variant of histone H3 named CENP-A, which is essential for the efficient binding of the kinetochore during mitosis. The kinetochore is a huge structure that attaches to centromeric DNA and mediates the interaction of chromosomes with the mitotic spindle and their movement to the spindle poles during mitosis [40]. Accurate segregation of chromosomes relies on the correct formation of the spindle apparatus.

CENP-A (also known as CENH3) is an essential protein that replaces histone H3 at centromeric DNA. This protein is highly divergent among different species but is functionally well conserved since the homologue protein of *S. cerevisiae,* Cse4 is able to complement human cells lacking CENP-A or *vice versa* [41]. In human cells, CENP-A is not assembled on to DNA just after DNA replication and CENP-A-containing nucleosomes are interspersed with canonical nucleosomes during replication of centromeres [42, 43]. This organization promotes the folding of centromeric chromatin into a unique structure during metaphase, in which all the nucleosomes containing CENP-A are facing the external side of chromosomes. This structure allows kinetochore assembly and microtubule attachment and favours sister chromatid cohesion [44]. Once chromosomes are separated, CENH3 is fully positioned on centromeric chromatin during the period between telophase and G1 in a process that is dependent on the transient incorporation of Mis18 and KNL2 in anaphase [45]. The incorporation of CENP-A to centromeric chromatin is mediated at least in part by the HJURP protein (Scm3 in *S. cerevisiae*) and is related to low levels of acetylation of the K16 residue of histone H4. Defects in the proper incorporation of this histone variant can lead to cell death, genetic instability and chromosome loss [46-48]. There is also a subset of specific proteins important to prevent the deposition of CENH3 containing nucleosomes out of the centromeric DNA. In *S. cerevisiae,* the ubiquitin E3 ligase Psh1 prevents the spread of Cse4 containing nucleosomes out of the centromere [49] [50]. The absence of both CAF1 and HIRA also leads to the presence of this type of nucleosomes in euchromatic regions in both *S. cerevisiae* and *S. pombe* and has been shown to cause genetic instability [51]. Finally, several papers point out that a proper homeostasis between H3 and CENH3 histones is important for the distribution of this centromeric variant and for efficient chromosome segregation [52, 53].

3. From gene to protein, histones are highly regulated

It is clear that there is a strong interdependency between DNA replication and chromatin reorganization. Nucleosomes are more than structural bricks for DNA, and require the modification of specific residues or the substitution of certain histone variants for others to maintain a correct epigenetic state. Once cells have decided to replicate their DNA, an increase in the abundance of histone proteins is required to pack the new genome that is about to be generated. Histone genes are among the most highly cell-cycle-regulated genes [54] because cells need to ensure a high demand of histones during replication, but must make sure that these levels are quickly down-regulated when replication slows down or is blocked, to avoid the deleterious effects of free histones on cell survival.

Canonical histone proteins can be regulated at transcriptional, post-transcriptional, translational and post-translational levels. The importance of each pathway on histone metabolism largely depends on the organism. In *S. cerevisiae* for example, the transcriptional regulation has been traditionally shown to play a major role in histone regulation, while in mammalian cells, post-transcriptional and translational mechanisms seem to be more important. Nevertheless, it is clear that all organisms try to produce histones exclusively during the replicative S-phase and more specifically only when replication is actively taking place.

3.1. Histones are regulated from the beginning: transcriptional regulation

Histone transcription is tightly regulated during the cell cycle. In some organisms like *S. cerevisiae*, transcription of histones can only be detected in late G1 and during DNA replication [55]. In higher eukaryotes, however, histones mRNAs can be found at all stages but increase as cells enter S-phase [56]. Expression of all canonical histones must be stoichiometric and several studies show that an imbalance between the different histone subtypes can be highly deleterious for the cell [52, 53, 57].

In metazoans, entry into S-phase increases the expression of replication-dependent histone genes three to five-fold [58]. Histone genes are clustered and each cluster normally contains at least one copy of the five canonical histones. Although transcription of all histones is carefully coordinated, no obvious common sequence element has been found at their promoters. Nevertheless, common elements can be found for some particular histone variants, like the Octamer-binding Transcription Factor (OTF1) for H2B promoters [59] or the Coding Region Activating Sequences (CRAS) in H2A, H3 and H4 genes [60]. Activation of histone gene transcription requires the Nuclear Protein Ataxia-Telangiectasia (NPAT), which is essential for S-phase progression [61]. This protein normally locates next to the *Histone locus bodies* and is phosphorylated by cyclin E-CDK2 at the beginning of S-Phase. Phosphorylation persists through S-phase and increases histone gene transcription [62, 63].

Transcriptional regulation of histones in *S. cerevisiae* is largely dependent on the integrity of the HIR complex. This complex is conserved from yeast to humans and has been shown to play a role in both of them in replication-independent chromatin assembly. In yeast, this complex is composed of the three subunits Hir1-3 and Hpc2. Deletion of any of the subunits leads to a de-repression of histones outside of S-phase [64]. Histone genes are grouped in 4 clusters, and each of them express simultaneously H2A/H2B or H3/H4 from a bidirectional promoter. These promoters contain upstream activating sequences (UAS) required for the recruitment of activators such as Spt10 and SBF [65]. Three of these four clusters also contain a negative regulatory site (NEG or CCR) able to maintain these genes in a repressed state in cell cycle phases outside of late G1 and S-phase and under replication stress conditions [64, 66, 67]. Deletion of the negative regulatory site is able to de-repress the *HTA1-HTB1* histone locus and allow expression outside of S-phase. The mechanism of repression is not completely understood but involves changes at the chromatin structure creating a repressive chromatin, which depends on the HIR complex, proteins Rtt106, Yta7 and Asf1, and the chromatin remodelling complex RSC. Two recent reports coming from the same group have shed some light on how repressive chromatin switches to active chromatin (in terms of transcription). The first one involves the degradation of Yta7 mediated by a phosphorylation event that involves Casein Kinase II (CKAII) and the cyclin-dependent kinase Cdc28. Degradation of Yta7 allows the efficient expression of histone mRNAs during S-phase through a mechanism that could involve transcription elongation efficiency [68]. The second report is related to the cell cycle regulation of Spt21, an activator of histone gene expression [69]. Spt21 protein levels outside S-phase are regulated by proteolysis, in a mechanism that depends on a complex formed by the Anaphase Promoting Complex (APC) with Cdh1 during G1, and on APC-Cdc20 during G2 and M (Brenda Andrews, EMBO transcription meeting 2012).

It has been recently shown that the HIR complex is conserved through evolution [70, 71] [55]. In humans, this complex is formed by three proteins: HIRA, Ubinuclein1 and Cabin1. The role described for HIRA in humans has been mostly associated with chromatin assembly of the transcriptional histone variant H3.3 in cooperation with ASF1 [72]. Nevertheless, several studies suggest that this complex could also play an important role in metazoan histone regulation. Ectopic over-expression of HIRA is able to repress histone gene transcription and block S-phase progression in human cells. This protein localizes to histone gene clusters in an immunofluorescence essay [73]. Cyclin E-CDK2 and cyclin A-CDK2 can phosphorylate HIRA, and this phosphorylation is inhibited by cyclin inhibitor p21, which has been shown to be important for repression of histone synthesis upon replication stress [74] [69]. HIRA could therefore be acting as a repressor of histone gene expression outside of S-phase regulated through phosphorylation by the cyclin E-CDK2. In this model, phosphorylation by the cyclin E-CDK2 could switch histone expression by activating NPAT and inactivating HIRA.

3.2. Once they are transcribed: post-transcriptional and translational regulation

Mammalian histone mRNAs lack introns and do not have a poly(A) tail as do most mRNAs. Instead, they contain a special 3′UTR sequence that forms a stem-loop structure [54]. Histone clusters localize to specific Cajal Bodies that are enriched in factors required for expression (NPAT) and maturation (U7 snRNA) of histone mRNA named *histone locus bodies* [75]. Maturation requires the formation of the 3′end through an endonucleolytic cleavage that has been shown to be important for transcription termination [76, 77]. Cleavage takes place between the stem-loop and the histone downstream element (HDE). The machinery involved in this process uses some common elements from the processing machinery of polyadenylated mRNAs but has also some specific components like SLBP, the Sm-like proteins (LSM1-11), the U7 snRNA and ZFP100. Additional information on maturation of histone mRNAs can be found in a nice review published some years ago [54]. The Stem Loop Binding Protein, SLBP, is one of the most important proteins for post-transcriptional and translational regulation of histone mRNAs and accompanies histone mRNA throughout its life.

SLBP is the only known cell cycle regulated protein of all the histone processing machinery. This protein starts accumulating during late G1 and is degraded at the end of S-phase by the phosphorylation of two threonine residues that target it for degradation [78]. There are three major roles for SLBP on histone regulation: 1. Allow an efficient cleavage during mRNA maturation 2. Facilitate circularization of histone mRNAs, required for their efficient translation by polyribosomes [79, 80] and 3. Increase histone mRNA stability preventing degradation of histone mRNAs by the 3′hExo [81]. Nevertheless, histone mRNA are still degraded when

SLBP is artificially present at constitutive levels at the end of S-phase [78] or when DNA replication is inhibited [82], indicating that although this protein has a major contribution to histone mRNA stability, it is not able to prevent degradation itself.

Canonical histone mRNAs in lower eukaryotes and plants are polyadenylated. The fact that these transcripts lack any known specific structure at their 3′end and have a short half-life has lead to the conclusion that regulation in these organisms mostly takes place at a transcriptional level. Nevertheless, there is quite a lot of recent evidence that strongly suggests the importance of the post-transcriptional regulation of histone mRNAs in S. cerevisiae. Several reports implicate some of the components of the exosome in the specific degradation of the H2B transcripts [83]. One year ago, Herrero and Moreno revealed the importance of the SM-like protein Lsm1 in histone mRNA degradation. Mutants lacking Lsm1 are sensitive to DNA damaging drugs and histone over-expression, and show a defect in histone mRNA degradation under replication stress conditions [84]. This protein is part of the Lsm1-7-Pat1 mRNA degradation complex, which has an important role in histone mRNA degradation under replication stress conditions in human cells. Lsm1-7-Pat1 has been shown to bind preferentially mRNAs carrying U-tracts in human cells, and oligoadenylated over polyadenylated mRNAs in yeast [85, 86]. Upon DNA replication arrest, histone mRNAs in human cells suffer an oligouridylation process acquiring a terminal oligo U-tract required for an efficient degradation by this complex [87]. Uridylation of mRNA has not been detected to date in S. cerevisiae but there is a recent report showing that the average length of the poly(A) tail of the yeast H2B histone mRNA is quite short compared to other transcripts. The length of this poly(A) is cell cycle-dependent and seems to decrease as cells exit G1 and progress through S-phase up to G2, when some of the transcripts completely lack a poly(A) tail [88]. This difference in length opens a possible explanation as to how the Lsm1-7 yeast complex preferentially recognizes yeast histone mRNAs over other transcripts to degrade them at the end of S-phase.

3.3. Last frontier of histone regulation: controling protein levels

In addition to the tight regulation of histone mRNA levels, a mechanism able to control histone protein levels was described some years ago [89]. To date, this pathway has only been described in the yeast S. cerevisiae and involves the action of the yeast homologue of CHK2, Rad53. Rad53 plays an important role in the DNA Damage Response and has been shown to be essential upon DNA damage or replication stress [1, 90]. Histone degradation involves the direct action of Rad53 along with the E2 ubiquitin ligases (UL) Ubc4 and Ubc5 and the E3 UL Tom1 [91]. This complex is able to degrade histones in a mechanism that involves tyrosine phosphorylation and poly-ubiquitylation, before their proteolysis by the proteasome. Histone degradation is independent on the central DNA damage checkpoint signal, since it does not depend on other kinases involved in the DDR like Mec1 (ATM) or Tel1 (ATR). Further studies in higher eukaryotes need to be done to confirm if this pathway is conserved in all eukaryotes

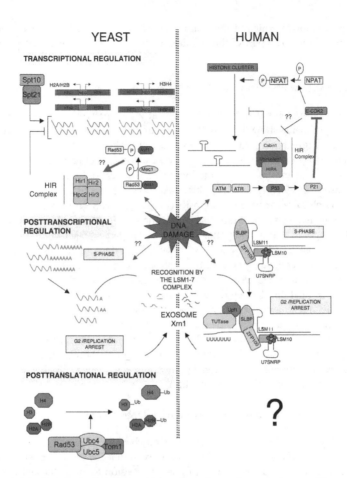

Figure 3. Regulation of histone levels in *S.cerevisiae* and *H.sapiens*. Different mechanisms able to control histone levels in *S.cerevisiae* and *H.sapiens*. Arrows normally indicate a positive effect on the pathway and straight lines a negative One. Interrogation marks are used when the protein/s involved in such process remain unknown or when the pathway has not been directly demonstrated. The big interrogation mark shown for post-translational regulation in the *H.sapiens* column, remarks that this pathway has not been demonstrated to date in human cells.

4. Histones: Enough to pack but not too much

Histone levels are regulated as soon as transcription of its mRNA starts. On top of the normal cell cycle regulation, additional mechanisms are able to block histone production when replication slows down or is completely blocked. Eukaryotic cells are unable to live without histones [92] and inhibition of histone deposition behind the replication fork blocks DNA synthesis and activates the DNA Damage Response (DDR) [35]. Eukaryotes seem to have evolved to a situation in which histones must not be free in the cell and DNA must not be free of histones. In this last part of the chapter, we will focus on how cells cope with situations that break this bidirectional relationship.

4.1. Harmful effects of free histones

Histones are basic proteins that can bind-specifically to negatively charged molecules. Reconstitution experiments show that a slight excess of histones over DNA is sufficient to promote chromatin aggregation, probably due to the neutralization of negative charged DNA. In yeast, high levels of histones increase chromosome loss and enhance DNA damage sensitivity [57, 89]. Defects in histone degradation during replication stress or DNA damage severely decreases cell viability [84]. Free histones show electrostatic interactions with some cellular macromolecules carrying the opposite charge such as RNA molecules. Additionally, an excess of free histones can saturate and inhibit the activity of some histone modifying enzymes, and change the expression pattern of almost 240 genes [93]. Two different studies in the yeasts *S. cerevisiae* and *Schizosaccharomyces pombe* have demonstrated the importance of a correct balance between histone H3 and the centromeric variant Cse4 (CENPA) for efficient chromosome segregation. H3 can compete with Cse4 in the assembly of centromeric chromatin and this competition largely depends on a correct balance between levels of H3 and H4 [52, 53]. Cells must therefore not only prevent the accumulation of free histones but also ensure a correct homeostasis between canonical and other histone variants. Once cells have decided to initiate replication, any problem that unbalances replication fork progression with histone levels can potentially lead to an increase in the abundance of free histones. In order to prevent this, there is an additional pathway linked to the DDR able to block histone synthesis under DNA replication stress conditions or replication fork arrest.

4.2. The DNA Damage Response (DDR): Coupling DNA and histone synthesis

DDR is probably one of the most well characterized checkpoints in the cell and is normally activated whenever a cell senses DNA damage. Activation leads to the sequential action of a cascade of kinases that block or delay cell cycle progression to allow the cell to correct the damage. If damage cannot be repaired, human cells enter the apoptosis program and die [94]. Proper functioning of this pathway is essential for genome integrity and mutations in most of the branches of this path are linked to cancer and other diseases. DDR is able to block cells at G1, S and G2/M [95]. In human cells, two kinases ATM and ATR (Tel1 and Mec1 respectively in *S. cerevisiae*) play a major role in the activation of the DDR. ATM has been directly involved in the activation of a mechanism that ultimately leads to repression of histone expression.

In human cells, activation of histone gene transcription requires NPAT phosphorylation by the cyclin E-CDK2 complex at the beginning of S-Phase. Activation of NPAT is essential for S-phase progression and histone expression. Repression of histone synthesis upon DNA damage requires the activation of ATM, which leads to the sequential activation of p53 and then p21. p21 is able to block the activity of the cyclin E-CDK2 complex. Inhibition of this complex leads to a progressive dephosphorylation of NPAT, which no longer localizes to histone clusters to activate transcription [96]. One interesting hypothesis that remains to be tested is if this cascade could also lead to histone repression by a change in the activity or location of HIRA, the human homologue of the HIR complex, at histone promoters (see previous paragraph about transcriptional regulation of histones). DNA damage also promotes post-transcriptional degradation of histone mRNAs. Treatment of cells with hydroxyurea (HU) increases oligouridylation of histone mRNAs in a process that depends on Upf1. Upf1 binds SLBP and helps to recruit a 3′ Terminal Uridylyl Transferase (TUT-ase) required for oligourydilation. These 3′ oligo(U) tails are recognized by the Lsm1–7 complex that triggers mRNA degradation through the exosome and Xrn1 [87]. How Upf1 is recruited to histone mRNAs upon DNA damage remains unknown.

Regulation of histone levels upon DNA damage in *S. cerevisiae* shows some common regulatory elements with human cells, and suggests the existence of a conserved mechanism. Post-transcriptional regulation also depends on the Lsm1-7 complex. It is not clear how this complex recognizes histone mRNAs but it could be related to the poly (A) tail-length (see post-transcriptional regulation of histones). Post-translational regulation by the Rad53 histone degradation pathway has not been directly addressed during the DDR, but taking into account the role of this protein in both pathways, it is reasonable to think that Rad53 could be important to destroy the population of translated histones when replication is halted. There are no NPAT homologues described in yeast and negative regulation during the DDR depends on the integrity of the HIR complex. The repressive structure formed to block transcription on histone promoters also requires Asf1 and Rtt106 among others. Although there is a lot of information about the formation of the repressive structure created at the promoter [97], the first steps by which DNA damage triggers histone repression remain largely unknown. There is some data nevertheless that suggest that Asf1 and Rad53 could play a role in this process.

Asf1 is able to form a very stable complex with Rad53. Upon activation of the DNA damage response, Mec1 phosphorylates Rad53 and this phosphorylation dissociates the stable Asf1-Rad53 complex. This mechanism has been linked to a possible connection between checkpoint activation and DNA repair since Asf1 plays a role in chromatin remodeling during DNA repair [98]. Rad53 can also be found in a hypophosphorylated form in normal conditions during G1, G2 and M, stages at which histone transcription is repressed. This phosphorylation seems to depend on Cdc28, the yeast functional homologue of human CDK1 and CDK2. Asf1 is able to co-immunoprecipitate with all the subunits of the HIR complex. This complex has been related to replication-independent nucleosome assembly and *in vitro* data prove that it is able to assembly nucleosomes to a DNA template [99]. Mutants lacking Asf1 have higher levels of histone mRNA and show defects in S-phase progression [100]. We have recently seen in our lab that mutants lacking the kinase activity of Rad53 also have enhanced levels of these mRNAs (unpublished results). Taking into account the close relationship that Asf1 plays with both Rad53 and the HIR complex, it is possible to think that the dissociation of Rad53 and Asf1 during DNA damage could be important for the efficient repression of histone transcription.

4.3. Generation of free histones in the cell

How can free histones be generated during a normal cell cycle? Taking into account the tight regulation of histone levels, such situations may seem unlikely. There are two scenarios in which it is possible to think that histone supply and DNA replication can be unbalanced during a normal cell cycle. In the first one, this situation could arise from differences between the rate of DNA synthesis and histone supply during replication. Early S-phase cells use more replication forks than late S-phase cells [101, 102] and lesions in DNA or replication stress also affect the speed of the replication fork [103-106]. The second scenario in which a cell can encounter free histones would take place during the G2 stage of the cell cycle. Given the importance of a balanced ratio between histone H3 and CENPA in chromosome segregation, once cells have finished replication, all free histones that are not positioned should be quickly degraded. It is possible to think that an imbalance between these two types of histones could sometimes take place in actively replicating cells and opens a simple explanation to why most cancer cells have a high incidence rate of chromosome loss [107].

4.4. Transcription as a source of free histones

Transcription of a chromatin template also requires nucleosomes to be disassembled and reassembled after the passage of RNA polymerase II (RNA Pol II). Outside of the S-phase, transcribed chromatin is probably the major potential source of free histones. These free histones could arise due to minor imbalances between histone supply and demand during chromatin reassembly. One very well described essential factor involved in RNA pol II transcription is the FACT complex [108, 109]. This complex is able to stimulate RNA Pol II-dependent transcription elongation through chromatin *in vitro* [110, 111] and also *in vivo* [112-114]. FACT is able to bind H3/H4 tetramers and H2A/H2B dimers [115, 116] and it has been shown that the integrity of at least one of its subunits, Spt16, is important for an efficient reassembly of the original H3 and H4 histones evicted during transcription [117]. Our group, in collaboration with others, demonstrated two years ago that dysfunction in chromatin reassembly during transcription due to defects in the Spt16 protein generates an accumulation of free histones in yeast. Combination of this mutant with a kinase dead version of Rad53 (*rad53K*^{227}*A*), unable to efficiently degrade histones, increases the accumulation of free histones and greatly impairs viability of this mutant in a checkpoint-independent way [118]. Deletion of one of the two-histone clusters for H2A-H2B expression is able to partially suppress the growth defect of this mutant and increasing H2A-H2B expression has the opposite effect. There is a strong correlation between histone levels and viability defects of the chromatin reassembly mutant of Spt16. This defect is not exclusive for Spt16, since Spt6, another chromatin remodeling factor involved in H3-H4 repositioning during transcription, also has a strong negative interaction with *rad53K227A*. Chromatin reassembly defects can lead to the generation of free histones evicted from chromatin during transcription, a new source of histones potentially toxic for the cell. Rad53 negatively interacts with many different proteins involved in chromatin-related processes and could have an important function in maintaining chromatin structure in yeast [119]. Some of these interactions are with factors that have only been exclusively involved to date in chromatin related processes during transcription [118, 119] suggesting that Rad53 could play an important role in the degradation of histones when chromatin is not correctly reassembled during transcription.

4.5. Can the state of chromatin influence the decision of cells to initiate DNA replication?

Histones are able to affect DNA replication right from the beginning; the state of the chromatin influences the timing and organization of origin firing. Replication fork progression also depends on the correct histone deposition behind the replication machinery, since defects in CAF1 lead to checkpoint activation and block cells in S-phase. The state of chromatin is able therefore to influence DNA replication. Work done in our lab, suggests that chromatin state might also influence the decision of cells to enter or not replication during the G1/S transition in the *S. cerevisiae*.

The commitment to a new round of cell division takes place towards the end of the G1 phase of the cell cycle in a process called START in yeast, and Restriction Point in mammals [120]. In yeast, this is the main regulatory event of the G1 phase of the cell cycle and involves an extensive transcriptional program driven by transcription factors SBF (Swi4-Swi6) and MBF (Mbp1-Swi6) [121, 122]. MBF and SBF activation depends on the cyclin/cyclin-dependent-kinase (CDK) complex Cln3-Cdc28. This complex phosphorylates Whi5, the negative regulator of START, thus promoting its release from SBF (Swi4-Swi6). Activation of MBF-dependent transcription by Cln3-Cdc28 acts through a mechanism independent of Whi5, involving the phosphorylation of Mbp1 [123]. Activation of these two complexes results in the accumulation of G1 (Cln1 and Cln2) and the early S-phase cyclins (Clb5 and Clb6), which promote in last term S-phase entry [124]. The kinase activity of Cln1,2-Cdc28 triggers the degradation of cyclin-dependent kinase inhibitor Sic1 which no longer inhibits the S phase-promoting complex Clb5,6-Cdc28 [125, 126].

FACT plays a role in maintaining the integrity of the chromatin structure during transcription [127-129] but has also been related to a G1/S cell cycle defect in yeast in a genetic screen to identify *cdc* (cell division cycle) mutants. This cell cycle defect had been linked initially to a general transcription defect of the three G1 cyclins Cln1-Cln3 [130] and later, to a possible important role of FACT in the transcription of *CLN1* and *CLN2* [131]. We recently described that this G1 defect is also due to a transcriptional downregulation of the cyclin Cln3 at the promoter level. Surprisingly, FACT seems not to be directly involved in the transcriptional regulation of this cyclin, since it is not recruited to the promoter at START when *CLN3* levels are maximal (D. Stillman unpublished results). One rather unexpected but interesting result is that this cell cycle defect shows a direct correlation with histone levels. Decreasing the H2A-H2B histone pool diminishes the cell cycle accumulation of this mutant while blocking the efficient degradation of histones has an additive effect. This defect is not exclusive for FACT mutants, since an Spt6 mutant also shows cell cycle defects at the G1/S transition [118]. Moreover, in yeast, a structural mutant of histone H4 in a region important for the interaction between the H3-H4 tetramer and the two H2A-H2B dimers completely mimics the cell cycle defects of the Spt16 mutant [132]. Defects in the chromatin structure seem to be connected somehow to the G1/S transition. Our group has speculated that cells might be able to sense the chromatin state before entering a new round of replication. This mechanism would act at least in part through a transcriptional repression of the cyclin Cln3 mRNA. Although our first results pointed out that free histones could be the signal that triggers this G1/S transition defect, new results obtained by our lab show that this regulation could be more complex and also involve the chromatin structure itself (unpublished results).

5. Conclusion

In eukaryotes and also some archaebacteria, DNA forms a nucleoprotein complex called chromatin, which allows extensive compaction of genomic DNA in the limited space of the nucleus. This traditional view of chromatin as simple building-bricks has substantially changed since the nucleosome hypothesis was proposed [133, 134]. Cells have evolved a unique and complex machinery to cope with the fact that most of the processes involving DNA are going to need to interact with and probably modify chromatin first. Chromatin acts as a new step of regulation and carries an epigenetic specific code that in some cases can be as important for the cell as the one contained on DNA. In addition, cells must also carefully balance the levels of histones during chromatin formation to avoid the generation of free histones, in order to prevent their deleterious effects.

Acknowledgments

We thank Akash Gunjan's and Vincent Geli's labs for our fruitful collaboration. We would also like to thank people in the lab for such a wonderful work environment and to all those people who are always there to make things easier. The Spanish Ministry of Education and Science (grant BFU2007-67575-C03-02/BMC), the Andalusian Government (grant P07-CVI02623) and the European Union (FEDER) have supported this work. D.M. was covered by a F.P.I. fellowship from the Regional Andalusian Government, and M. M-H. and L.D-R., by fellowships from the Spanish Ministry of Education and Science.

Author details

Douglas Maya*, Macarena Morillo-Huesca, Lidia Delgado Ramos, Sebastián Chávez and Mari-Cruz Muñoz-Centeno

*Address all correspondence to: dmaya@us.es

Department of Genetics, Faculty of Biology, University of Seville, Spain

References

[1] Jones RM, Petermann E. Replication fork dynamics and the DNA damage response. Biochem J. 2012 Apr 1;443(1):13-26.

[2] Gunjan A, Paik J, Verreault A. Regulation of histone synthesis and nucleosome assembly. Biochimie. 2005 Jul;87(7):625-635.

[3] Alabert C, Groth A. Chromatin replication and epigenome maintenance. Nat Rev Mol Cell Biol. 2012 Mar;13(3):153-167.

[4] Mechali M. Eukaryotic DNA replication origins: many choices for appropriate answers. Nat Rev Mol Cell Biol. 2010 Oct;11(10):728-738.

[5] Xu J, Yanagisawa Y, Tsankov AM, Hart C, Aoki K, Kommajosyula N, et al. Genome-wide identification and characterization of replication origins by deep sequencing. Genome Biol. 2012 Apr 24;13(4):R27.

[6] Gao F, Luo H, Zhang CT. DeOri: a database of eukaryotic DNA replication origins. Bioinformatics. 2012 Jun 1;28(11):1551-1552.

[7] Sclafani RA, Holzen TM. Cell cycle regulation of DNA replication. Annu Rev Genet. 2007;41:237-280.

[8] Dohrmann PR, Sclafani RA. Novel role for checkpoint Rad53 protein kinase in the initiation of chromosomal DNA replication in Saccharomyces cerevisiae. Genetics. 2006 Sep;174(1):87-99.

[9] Aggarwal BD, Calvi BR. Chromatin regulates origin activity in Drosophila follicle cells. Nature. 2004 Jul 15;430(6997):372-376.

[10] Tardat M, Brustel J, Kirsh O, Lefevbre C, Callanan M, Sardet C, et al. The histone H4 Lys 20 methyltransferase PR-Set7 regulates replication origins in mammalian cells. Nat Cell Biol. 2010 Nov;12(11):1086-1093.

[11] Kouzarides T. Chromatin modifications and their function. Cell. 2007 Feb 23;128(4): 693-705.

[12] Miotto B, Struhl K. HBO1 histone acetylase is a coactivator of the replication licensing factor Cdt1. Genes Dev. 2008 Oct 1;22(19):2633-2638.

[13] Miotto B, Struhl K. HBO1 histone acetylase activity is essential for DNA replication licensing and inhibited by Geminin. Mol Cell. 2010 Jan 15;37(1):57-66.

[14] Iizuka M, Matsui T, Takisawa H, Smith MM. Regulation of replication licensing by acetyltransferase Hbo1. Mol Cell Biol. 2006 Feb;26(3):1098-1108.

[15] Remus D, Diffley JF. Eukaryotic DNA replication control: lock and load, then fire. Curr Opin Cell Biol. 2009 Dec;21(6):771-777.

[16] MacAlpine HK, Gordan R, Powell SK, Hartemink AJ, MacAlpine DM. Drosophila ORC localizes to open chromatin and marks sites of cohesin complex loading. Genome Res. 2010 Feb;20(2):201-211.

[17] Guillou E, Ibarra A, Coulon V, Casado-Vela J, Rico D, Casal I, et al. Cohesin organizes chromatin loops at DNA replication factories. Genes Dev. 2010 Dec 15;24(24): 2812-2822.

[18] Courbet S, Gay S, Arnoult N, Wronka G, Anglana M, Brison O, et al. Replication fork movement sets chromatin loop size and origin choice in mammalian cells. Nature. 2008 Sep 25;455(7212):557-560.

[19] Gilbert DM, Takebayashi SI, Ryba T, Lu J, Pope BD, Wilson KA, et al. Space and time in the nucleus: developmental control of replication timing and chromosome architecture. Cold Spring Harb Symp Quant Biol. 2010;75:143-153.

[20] Schwaiger M, Stadler MB, Bell O, Kohler H, Oakeley EJ, Schubeler D. Chromatin state marks cell-type- and gender-specific replication of the Drosophila genome. Genes Dev. 2009 Mar 1;23(5):589-601.

[21] Sogo JM, Stahl H, Koller T, Knippers R. Structure of replicating simian virus 40 minichromosomes. The replication fork, core histone segregation and terminal structures. J Mol Biol. 1986 May 5;189(1):189-204.

[22] Gasser R, Koller T, Sogo JM. The stability of nucleosomes at the replication fork. J Mol Biol. 1996 May 3;258(2):224-239.

[23] Kuipers MA, Stasevich TJ, Sasaki T, Wilson KA, Hazelwood KL, McNally JG, et al. Highly stable loading of Mcm proteins onto chromatin in living cells requires replication to unload. J Cell Biol. 2011 Jan 10;192(1):29-41.

[24] Alexandrow MG, Hamlin JL. Chromatin decondensation in S-phase involves recruitment of Cdk2 by Cdc45 and histone H1 phosphorylation. J Cell Biol. 2005 Mar 14;168(6):875-886.

[25] Belotserkovskaya R, Oh S, Bondarenko VA, Orphanides G, Studitsky VM, Reinberg D. FACT facilitates transcription-dependent nucleosome alteration. Science. 2003 Aug 22;301(5636):1090-1093.

[26] Xin H, Takahata S, Blanksma M, McCullough L, Stillman DJ, Formosa T. yFACT induces global accessibility of nucleosomal DNA without H2A-H2B displacement. Mol Cell. 2009 Aug 14;35(3):365-376.

[27] Hertel L, De Andrea M, Bellomo G, Santoro P, Landolfo S, Gariglio M. The HMG protein T160 colocalizes with DNA replication foci and is down-regulated during cell differentiation. Exp Cell Res. 1999 Aug 1;250(2):313-328.

[28] Tan BC, Chien CT, Hirose S, Lee SC. Functional cooperation between FACT and MCM helicase facilitates initiation of chromatin DNA replication. EMBO J. 2006 Sep 6;25(17):3975-3985.

[29] Ito T, Bulger M, Kobayashi R, Kadonaga JT. Drosophila NAP-1 is a core histone chaperone that functions in ATP-facilitated assembly of regularly spaced nucleosomal arrays. Mol Cell Biol. 1996 Jun;16(6):3112-3124.

[30] Lorch Y, Maier-Davis B, Kornberg RD. Chromatin remodeling by nucleosome disassembly in vitro. Proc Natl Acad Sci U S A. 2006 Feb 28;103(9):3090-3093.

[31] Xu M, Long C, Chen X, Huang C, Chen S, Zhu B. Partitioning of histone H3-H4 tetramers during DNA replication-dependent chromatin assembly. Science. 2010 Apr 2;328(5974):94-98.

[32] Franco AA, Lam WM, Burgers PM, Kaufman PD. Histone deposition protein Asf1 maintains DNA replisome integrity and interacts with replication factor C. Genes Dev. 2005 Jun 1;19(11):1365-1375.

[33] Groth A, Corpet A, Cook AJ, Roche D, Bartek J, Lukas J, et al. Regulation of replication fork progression through histone supply and demand. Science. 2007 Dec 21;318(5858):1928-1931.

[34] Hoek M, Stillman B. Chromatin assembly factor 1 is essential and couples chromatin assembly to DNA replication in vivo. Proc Natl Acad Sci U S A. 2003 Oct 14;100(21): 12183-12188.

[35] Ye X, Franco AA, Santos H, Nelson DM, Kaufman PD, Adams PD. Defective S phase chromatin assembly causes DNA damage, activation of the S phase checkpoint, and S phase arrest. Mol Cell. 2003 Feb;11(2):341-351.

[36] Finn RM, Browne K, Hodgson KC, Ausio J. sNASP, a histone H1-specific eukaryotic chaperone dimer that facilitates chromatin assembly. Biophys J. 2008 Aug;95(3): 1314-1325.

[37] Smith DJ, Whitehouse I. Intrinsic coupling of lagging-strand synthesis to chromatin assembly. Nature. 2012 Mar 22;483(7390):434-438.

[38] Pesavento JJ, Yang H, Kelleher NL, Mizzen CA. Certain and progressive methylation of histone H4 at lysine 20 during the cell cycle. Mol Cell Biol. 2008 Jan;28(1):468-486.

[39] Lanzuolo C, Lo Sardo F, Diamantini A, Orlando V. PcG complexes set the stage for epigenetic inheritance of gene silencing in early S phase before replication. PLoS Genet. 2011 Nov;7(11):e1002370.

[40] Westermann S, Drubin DG, Barnes G. Structures and functions of yeast kinetochore complexes. Annu Rev Biochem. 2007;76:563-591.

[41] Wieland G, Orthaus S, Ohndorf S, Diekmann S, Hemmerich P. Functional complementation of human centromere protein A (CENP-A) by Cse4p from Saccharomyces cerevisiae. Mol Cell Biol. 2004 Aug;24(15):6620-6630.

[42] Jansen LE, Black BE, Foltz DR, Cleveland DW. Propagation of centromeric chromatin requires exit from mitosis. J Cell Biol. 2007 Mar 12;176(6):795-805.

[43] Schuh M, Lehner CF, Heidmann S. Incorporation of Drosophila CID/CENP-A and CENP-C into centromeres during early embryonic anaphase. Curr Biol. 2007 Feb 6;17(3):237-243.

[44] Black BE, Bassett EA. The histone variant CENP-A and centromere specification. Curr Opin Cell Biol. 2008 Feb;20(1):91-100.

[45] Fujita Y, Hayashi T, Kiyomitsu T, Toyoda Y, Kokubu A, Obuse C, et al. Priming of centromere for CENP-A recruitment by human hMis18alpha, hMis18beta, and M18BP1. Dev Cell. 2007 Jan;12(1):17-30.

[46] Choy JS, Mishra PK, Au WC, Basrai MA. Insights into assembly and regulation of centromeric chromatin in Saccharomyces cerevisiae. Biochim Biophys Acta. 2012 Jul; 1819(7):776-783.

[47] Stoler S, Rogers K, Weitze S, Morey L, Fitzgerald-Hayes M, Baker RE. Scm3, an essential Saccharomyces cerevisiae centromere protein required for G2/M progression and Cse4 localization. Proc Natl Acad Sci U S A. 2007 Jun 19;104(25):10571-10576.

[48] Dunleavy EM, Roche D, Tagami H, Lacoste N, Ray-Gallet D, Nakamura Y, et al. HJURP is a cell-cycle-dependent maintenance and deposition factor of CENP-A at centromeres. Cell. 2009 May 1;137(3):485-497.

[49] Hewawasam G, Shivaraju M, Mattingly M, Venkatesh S, Martin-Brown S, Florens L, et al. Psh1 is an E3 ubiquitin ligase that targets the centromeric histone variant Cse4. Mol Cell. 2010 Nov 12;40(3):444-454.

[50] Ranjitkar P, Press MO, Yi X, Baker R, MacCoss MJ, Biggins S. An E3 ubiquitin ligase prevents ectopic localization of the centromeric histone H3 variant via the centromere targeting domain. Mol Cell. 2010 Nov 12;40(3):455-464.

[51] Lopes da Rosa J, Holik J, Green EM, Rando OJ, Kaufman PD. Overlapping regulation of CenH3 localization and histone H3 turnover by CAF-1 and HIR proteins in Saccharomyces cerevisiae. Genetics. 2011 Jan;187(1):9-19.

[52] Castillo AG, Mellone BG, Partridge JF, Richardson W, Hamilton GL, Allshire RC, et al. Plasticity of fission yeast CENP-A chromatin driven by relative levels of histone H3 and H4. PLoS Genet. 2007 Jul;3(7):e121.

[53] Au WC, Crisp MJ, DeLuca SZ, Rando OJ, Basrai MA. Altered dosage and mislocalization of histone H3 and Cse4p lead to chromosome loss in Saccharomyces cerevisiae. Genetics. 2008 May;179(1):263-275.

[54] Marzluff WF, Wagner EJ, Duronio RJ. Metabolism and regulation of canonical histone mRNAs: life without a poly(A) tail. Nat Rev Genet. 2008 Nov;9(11):843-854.

[55] Amin AD, Vishnoi N, Prochasson P. A global requirement for the HIR complex in the assembly of chromatin. Biochim Biophys Acta. 2012 Mar;1819(3-4):264-276.

[56] Jaeger S, Barends S, Giege R, Eriani G, Martin F. Expression of metazoan replication-dependent histone genes. Biochimie. 2005 Sep-Oct;87(9-10):827-834.

[57] Meeks-Wagner D, Hartwell LH. Normal stoichiometry of histone dimer sets is necessary for high fidelity of mitotic chromosome transmission. Cell. 1986 Jan 17;44(1): 43-52.

[58] Marzluff WF, Duronio RJ. Histone mRNA expression: multiple levels of cell cycle regulation and important developmental consequences. Curr Opin Cell Biol. 2002 Dec;14(6):692-699.

[59] Fletcher C, Heintz N, Roeder RG. Purification and characterization of OTF-1, a transcription factor regulating cell cycle expression of a human histone H2b gene. Cell. 1987 Dec 4;51(5):773-781.

[60] Kaludov NK, Pabon-Pena L, Hurt MM. Identification of a second conserved element within the coding sequence of a mouse H3 histone gene that interacts with nuclear factors and is necessary for normal expression. Nucleic Acids Res. 1996 Feb 1;24(3): 523-531.

[61] Ye X, Wei Y, Nalepa G, Harper JW. The cyclin E/Cdk2 substrate p220(NPAT) is required for S-phase entry, histone gene expression, and Cajal body maintenance in human somatic cells. Mol Cell Biol. 2003 Dec;23(23):8586-8600.

[62] Ma T, Van Tine BA, Wei Y, Garrett MD, Nelson D, Adams PD, et al. Cell cycle-regulated phosphorylation of p220(NPAT) by cyclin E/Cdk2 in Cajal bodies promotes histone gene transcription. Genes Dev. 2000 Sep 15;14(18):2298-2313.

[63] Zhao J, Kennedy BK, Lawrence BD, Barbie DA, Matera AG, Fletcher JA, et al. NPAT links cyclin E-Cdk2 to the regulation of replication-dependent histone gene transcription. Genes Dev. 2000 Sep 15;14(18):2283-2297.

[64] Spector MS, Raff A, DeSilva H, Lee K, Osley MA. Hir1p and Hir2p function as transcriptional corepressors to regulate histone gene transcription in the Saccharomyces cerevisiae cell cycle. Mol Cell Biol. 1997 Feb;17(2):545-552.

[65] Eriksson PR, Ganguli D, Clark DJ. Spt10 and Swi4 control the timing of histone H2A/H2B gene activation in budding yeast. Mol Cell Biol. 2011 Feb;31(3):557-572.

[66] Osley MA, Gould J, Kim S, Kane MY, Hereford L. Identification of sequences in a yeast histone promoter involved in periodic transcription. Cell. 1986 May 23;45(4): 537-544.

[67] Freeman KB, Karns LR, Lutz KA, Smith MM. Histone H3 transcription in Saccharomyces cerevisiae is controlled by multiple cell cycle activation sites and a constitutive negative regulatory element. Mol Cell Biol. 1992 Dec;12(12):5455-5463.

[68] Kurat CF, Lambert JP, van Dyk D, Tsui K, van Bakel H, Kaluarachchi S, et al. Restriction of histone gene transcription to S phase by phosphorylation of a chromatin boundary protein. Genes Dev. 2011 Dec 1;25(23):2489-2501.

[69] Dollard C, Ricupero-Hovasse SL, Natsoulis G, Boeke JD, Winston F. SPT10 and SPT21 are required for transcription of particular histone genes in Saccharomyces cerevisiae. Mol Cell Biol. 1994 Aug;14(8):5223-5228.

[70] Balaji S, Iyer LM, Aravind L. HPC2 and ubinuclein define a novel family of histone chaperones conserved throughout eukaryotes. Mol Biosyst. 2009 Mar;5(3):269-275.

[71] Banumathy G, Somaiah N, Zhang R, Tang Y, Hoffmann J, Andrake M, et al. Human UBN1 is an ortholog of yeast Hpc2p and has an essential role in the HIRA/ASF1a chromatin-remodeling pathway in senescent cells. Mol Cell Biol. 2009 Feb;29(3): 758-770.

[72] Tagami H, Ray-Gallet D, Almouzni G, Nakatani Y. Histone H3.1 and H3.3 complexes mediate nucleosome assembly pathways dependent or independent of DNA synthesis. Cell. 2004 Jan 9;116(1):51-61.

[73] Nelson DM, Ye X, Hall C, Santos H, Ma T, Kao GD, et al. Coupling of DNA synthesis and histone synthesis in S phase independent of cyclin/cdk2 activity. Mol Cell Biol. 2002 Nov;22(21):7459-7472.

[74] Hall C, Nelson DM, Ye X, Baker K, DeCaprio JA, Seeholzer S, et al. HIRA, the human homologue of yeast Hir1p and Hir2p, is a novel cyclin-cdk2 substrate whose expression blocks S-phase progression. Mol Cell Biol. 2001 Mar;21(5):1854-1865.

[75] Nizami Z, Deryusheva S, Gall JG. The Cajal body and histone locus body. Cold Spring Harb Perspect Biol. 2010 Jul;2(7):a000653.

[76] Chodchoy N, Pandey NB, Marzluff WF. An intact histone 3'-processing site is required for transcription termination in a mouse histone H2a gene. Mol Cell Biol. 1991 Jan;11(1):497-509.

[77] Gu X, Marzluff WF. 3' Processing and termination of mouse histone transcripts synthesized in vitro by RNA polymerase II. Nucleic Acids Res. 1996 Oct 1;24(19): 3797-3805.

[78] Zheng L, Dominski Z, Yang XC, Elms P, Raska CS, Borchers CH, et al. Phosphorylation of stem-loop binding protein (SLBP) on two threonines triggers degradation of SLBP, the sole cell cycle-regulated factor required for regulation of histone mRNA processing, at the end of S phase. Mol Cell Biol. 2003 Mar;23(5):1590-1601.

[79] Sanchez R, Marzluff WF. The stem-loop binding protein is required for efficient translation of histone mRNA in vivo and in vitro. Mol Cell Biol. 2002 Oct;22(20): 7093-7104.

[80] Gorgoni B, Andrews S, Schaller A, Schumperli D, Gray NK, Muller B. The stem-loop binding protein stimulates histone translation at an early step in the initiation pathway. RNA. 2005 Jul;11(7):1030-1042.

[81] Huang Y, Gattoni R, Stevenin J, Steitz JA. SR splicing factors serve as adapter proteins for TAP-dependent mRNA export. Mol Cell. 2003 Mar;11(3):837-843.

[82] Whitfield ML, Kaygun H, Erkmann JA, Townley-Tilson WH, Dominski Z, Marzluff WF. SLBP is associated with histone mRNA on polyribosomes as a component of the histone mRNP. Nucleic Acids Res. 2004;32(16):4833-4842.

[83] Reis CC, Campbell JL. Contribution of Trf4/5 and the nuclear exosome to genome stability through regulation of histone mRNA levels in Saccharomyces cerevisiae. Genetics. 2007 Mar;175(3):993-1010.

[84] Herrero AB, Moreno S. Lsm1 promotes genomic stability by controlling histone mRNA decay. EMBO J. 2011 May 18;30(10):2008-2018.

[85] Chowdhury A, Mukhopadhyay J, Tharun S. The decapping activator Lsm1p-7p-Pat1p complex has the intrinsic ability to distinguish between oligoadenylated and polyadenylated RNAs. RNA. 2007 Jul;13(7):998-1016.

[86] Tharun S, He W, Mayes AE, Lennertz P, Beggs JD, Parker R. Yeast Sm-like proteins function in mRNA decapping and decay. Nature. 2000 Mar 30;404(6777):515-518.

[87] Mullen TE, Marzluff WF. Degradation of histone mRNA requires oligouridylation followed by decapping and simultaneous degradation of the mRNA both 5' to 3' and 3' to 5'. Genes Dev. 2008 Jan 1;22(1):50-65.

[88] Beggs S, James TC, Bond U. The PolyA tail length of yeast histone mRNAs varies during the cell cycle and is influenced by Sen1p and Rrp6p. Nucleic Acids Res. 2012 Mar;40(6):2700-2711.

[89] Gunjan A, Verreault A. A Rad53 kinase-dependent surveillance mechanism that regulates histone protein levels in S. cerevisiae. Cell. 2003 Nov 26;115(5):537-549.

[90] Navadgi-Patil VM, Burgers PM. Cell-cycle-specific activators of the Mec1/ATR checkpoint kinase. Biochem Soc Trans. 2011 Apr;39(2):600-605.

[91] Singh RK, Kabbaj MH, Paik J, Gunjan A. Histone levels are regulated by phosphorylation and ubiquitylation-dependent proteolysis. Nat Cell Biol. 2009 Aug;11(8): 925-933.

[92] Gunjan A, Paik J, Verreault A. The emergence of regulated histone proteolysis. Curr Opin Genet Dev. 2006 Apr;16(2):112-118.

[93] Singh RK, Liang D, Gajjalaiahvari UR, Kabbaj MH, Paik J, Gunjan A. Excess histone levels mediate cytotoxicity via multiple mechanisms. Cell Cycle. 2010 Oct 15;9(20): 4236-4244.

[94] Hanel W, Moll UM. Links between mutant p53 and genomic instability. J Cell Biochem. 2012 Feb;113(2):433-439.

[95] Ubezio P, Lupi M, Branduardi D, Cappella P, Cavallini E, Colombo V, et al. Quantitative assessment of the complex dynamics of G1, S, and G2-M checkpoint activities. Cancer Res. 2009 Jun 15;69(12):5234-5240.

[96]   Su C, Gao G, Schneider S, Helt C, Weiss C, O'Reilly MA, et al. DNA damage induces downregulation of histone gene expression through the G1 checkpoint pathway. EMBO J. 2004 Mar 10;23(5):1133-1143.

[97]   Eriksson PR, Ganguli D, Nagarajavel V, Clark DJ. Regulation of histone gene expression in budding yeast. Genetics. 2012 May;191(1):7-20.

[98]   Emili A, Schieltz DM, Yates JR, 3rd, Hartwell LH. Dynamic interaction of DNA damage checkpoint protein Rad53 with chromatin assembly factor Asf1. Mol Cell. 2001 Jan;7(1):13-20.

[99]   Green EM, Antczak AJ, Bailey AO, Franco AA, Wu KJ, Yates JR, 3rd, et al. Replication-independent histone deposition by the HIR complex and Asf1. Curr Biol. 2005 Nov 22;15(22):2044-2049.

[100] Sutton A, Bucaria J, Osley MA, Sternglanz R. Yeast ASF1 protein is required for cell cycle regulation of histone gene transcription. Genetics. 2001 Jun;158(2):587-596.

[101] Berezney R, Dubey DD, Huberman JA. Heterogeneity of eukaryotic replicons, replicon clusters, and replication foci. Chromosoma. 2000 Mar;108(8):471-484.

[102] Leonhardt H, Rahn HP, Weinzierl P, Sporbert A, Cremer T, Zink D, et al. Dynamics of DNA replication factories in living cells. J Cell Biol. 2000 Apr 17;149(2):271-280.

[103] Paulovich AG, Hartwell LH. A checkpoint regulates the rate of progression through S phase in S. cerevisiae in response to DNA damage. Cell. 1995 Sep 8;82(5):841-847.

[104] Santocanale C, Diffley JF. A Mec1- and Rad53-dependent checkpoint controls late-firing origins of DNA replication. Nature. 1998 Oct 8;395(6702):615-618.

[105] Feijoo C, Hall-Jackson C, Wu R, Jenkins D, Leitch J, Gilbert DM, et al. Activation of mammalian Chk1 during DNA replication arrest: a role for Chk1 in the intra-S phase checkpoint monitoring replication origin firing. J Cell Biol. 2001 Sep 3;154(5):913-923.

[106] Tercero JA, Diffley JF. Regulation of DNA replication fork progression through damaged DNA by the Mec1/Rad53 checkpoint. Nature. 2001 Aug 2;412(6846):553-557.

[107] Lengauer C, Kinzler KW, Vogelstein B. Genetic instabilities in human cancers. Nature. 1998 Dec 17;396(6712):643-649.

[108] Reinberg D, Sims RJ, 3rd. de FACTo nucleosome dynamics. J Biol Chem. 2006 Aug 18;281(33):23297-23301.

[109] Formosa T. FACT and the reorganized nucleosome. Mol Biosyst. 2008 Nov;4(11): 1085-1093.

[110] Orphanides G, LeRoy G, Chang CH, Luse DS, Reinberg D. FACT, a factor that facilitates transcript elongation through nucleosomes. Cell. 1998 Jan 9;92(1):105-116.

[111] Pavri R, Zhu B, Li G, Trojer P, Mandal S, Shilatifard A, et al. Histone H2B monoubiquitination functions cooperatively with FACT to regulate elongation by RNA polymerase II. Cell. 2006 May 19;125(4):703-717.

[112]   Mason PB, Struhl K. The FACT complex travels with elongating RNA polymerase II and is important for the fidelity of transcriptional initiation in vivo. Mol Cell Biol. 2003 Nov;23(22):8323-8333.

[113]   Saunders A, Werner J, Andrulis ED, Nakayama T, Hirose S, Reinberg D, et al. Tracking FACT and the RNA polymerase II elongation complex through chromatin in vivo. Science. 2003 Aug 22;301(5636):1094-1096.

[114]   Jimeno-Gonzalez S, Gomez-Herreros F, Alepuz PM, Chavez S. A gene-specific requirement for FACT during transcription is related to the chromatin organization of the transcribed region. Mol Cell Biol. 2006 Dec;26(23):8710-8721.

[115]   Stuwe T, Hothorn M, Lejeune E, Rybin V, Bortfeld M, Scheffzek K, et al. The FACT Spt16 "peptidase" domain is a histone H3-H4 binding module. Proc Natl Acad Sci U S A. 2008 Jul 1;105(26):8884-8889.

[116]   VanDemark AP, Xin H, McCullough L, Rawlins R, Bentley S, Heroux A, et al. Structural and functional analysis of the Spt16p N-terminal domain reveals overlapping roles of yFACT subunits. J Biol Chem. 2008 Feb 22;283(8):5058-5068.

[117]   Jamai A, Puglisi A, Strubin M. Histone chaperone spt16 promotes redeposition of the original h3-h4 histones evicted by elongating RNA polymerase. Mol Cell. 2009 Aug 14;35(3):377-383.

[118]   Morillo-Huesca M, Maya D, Munoz-Centeno MC, Singh RK, Oreal V, Reddy GU, et al. FACT prevents the accumulation of free histones evicted from transcribed chromatin and a subsequent cell cycle delay in G1. PLoS Genet. 2010 May;6(5):e1000964.

[119]   Pan X, Ye P, Yuan DS, Wang X, Bader JS, Boeke JD. A DNA integrity network in the yeast Saccharomyces cerevisiae. Cell. 2006 Mar 10;124(5):1069-1081.

[120]   Hartwell LH, Kastan MB. Cell cycle control and cancer. Science. 1994 Dec 16;266(5192):1821-1828.

[121]   Costanzo M, Nishikawa JL, Tang X, Millman JS, Schub O, Breitkreuz K, et al. CDK activity antagonizes Whi5, an inhibitor of G1/S transcription in yeast. Cell. 2004 Jun 25;117(7):899-913.

[122]   de Bruin RA, McDonald WH, Kalashnikova TI, Yates J, 3rd, Wittenberg C. Cln3 activates G1-specific transcription via phosphorylation of the SBF bound repressor Whi5. Cell. 2004 Jun 25;117(7):887-898.

[123]   Ferrezuelo F, Colomina N, Futcher B, Aldea M. The transcriptional network activated by Cln3 cyclin at the G1-to-S transition of the yeast cell cycle. Genome Biol. 2010;11(6):R67.

[124]   Wittenberg C. Cell cycle: cyclin guides the way. Nature. 2005 Mar 3;434(7029):34-35.

[125]   Schneider BL, Yang QH, Futcher AB. Linkage of replication to start by the Cdk inhibitor Sic1. Science. 1996 Apr 26;272(5261):560-562.

[126] Schwob E, Bohm T, Mendenhall MD, Nasmyth K. The B-type cyclin kinase inhibitor p40SIC1 controls the G1 to S transition in S. cerevisiae. Cell. 1994 Oct 21;79(2): 233-244.

[127] Formosa T, Ruone S, Adams MD, Olsen AE, Eriksson P, Yu Y, et al. Defects in SPT16 or POB3 (yFACT) in Saccharomyces cerevisiae cause dependence on the Hir/Hpc pathway: polymerase passage may degrade chromatin structure. Genetics. 2002 Dec; 162(4):1557-1571.

[128] Kaplan CD, Laprade L, Winston F. Transcription elongation factors repress transcription initiation from cryptic sites. Science. 2003 Aug 22;301(5636):1096-1099.

[129] Cheung V, Chua G, Batada NN, Landry CR, Michnick SW, Hughes TR, et al. Chromatin- and transcription-related factors repress transcription from within coding regions throughout the Saccharomyces cerevisiae genome. PLoS Biol. 2008 Nov 11;6(11):e277.

[130] Lycan D, Mikesell G, Bunger M, Breeden L. Differential effects of Cdc68 on cell cycle-regulated promoters in Saccharomyces cerevisiae. Mol Cell Biol. 1994 Nov;14(11): 7455-7465.

[131] Takahata S, Yu Y, Stillman DJ. The E2F functional analogue SBF recruits the Rpd3(L) HDAC, via Whi5 and Stb1, and the FACT chromatin reorganizer, to yeast G1 cyclin promoters. EMBO J. 2009 Nov 4;28(21):3378-3389.

[132] Santisteban MS, Arents G, Moudrianakis EN, Smith MM. Histone octamer function in vivo: mutations in the dimer-tetramer interfaces disrupt both gene activation and repression. EMBO J. 1997 May 1;16(9):2493-2506.

[133] Olins AL, Olins DE. Spheroid chromatin units (v bodies). Science. 1974 Jan 25;183(4122):330-332.

[134] Kornberg RD. Chromatin structure: a repeating unit of histones and DNA. Science. 1974 May 24;184(4139):868-871.

Replicating – DNA in the Refractory Chromatin Environment

Angélique Galvani and Christophe Thiriet

Additional information is available at the end of the chapter

1. Introduction

The replication of DNA is a process found throughout the prokaryotic and the eukaryotic kingdoms. Although the basic aim of this process is the duplication of the genetic information, the mechanisms leading to replication are different in prokaryotes and in eukaryotes. A major divergence between the two kingdoms corresponds to the nature of the substrate of the replication process [1]. Indeed, while the genetic information in prokaryotic cell is recovered in the nucleoid, the eukaryotic genome is found in the nucleus and the genetic material is associated with proteins. The tight interaction of the DNA molecule with proteins forms the chromatin, and for replication as well as for the other cellular processes that require the access to the genetic material, the chromatin is the actual substrate [2]. This organization of the eukaryotic genome in chromatin generates additional constraints to enzymatic activities. Therefore, it is required for the replication machinery to over-rule the refractory environment of chromatin.

Although the arrangement of the genetic material with proteins is an inhibitory environment, it is also required for packaging the molecule of DNA within the confined nuclear volume and for organizing the genome. Therefore, defects in the genetic material packaging affect genome stability and cell viability. Importantly, as replication results in the doubling of DNA, it is required for the cell to synthesize DNA-associated proteins and to form chromatin. This process known as replication-coupled chromatin assembly implies the copy of the epigenetic information carried by the histone proteins [3].

In the present chapter, we define the general features of chromatin, primarily on the basis of the fundamental sub-unit, the nucleosome, and the constraints that this structure generates for creating a refractory environment to replication. In addition to the view of the single nucleosome, as chromatin can be viewed as a polymer of nucleosomes which are highly ordered, the impediment of the replication machinery induced by higher chromatin order is discussed. Although replication activity should be inhibited by the chromatin, we review the mechanisms developed by eukaryotic cells to over-rule this non-permissive environment. Genetic experiments have shown that chromatin structure is essential for cell viability. We review the data providing evidence that the genome stability is, at least partly, inherent to chromatin assembly during replication, and the histone requirement in this process.

2. Chromatin: From the nucleosome sub-unit to the higher order structure

The basic chromatin sub-unit is the nucleosome, which is composed of the association of histone proteins with DNA [4]. The histone proteins are the most abundant nuclear proteins and are divided in four classes, H2A, H2B, H3 and H4, respectively. We distinguish in the histone protein two regions, the histone fold domain which is involved in the histone-histone and histone-DNA interactions, and the histone tail domain located at the N-terminal part of the protein, which is unstructured and extends out of the nucleosome [2, 5](Figure 1A). The association of the histones via their fold domain is highly conserved throughout the eukaryotic kingdom. Indeed, H3 is always associated with H4 and H2A with H2B forming therefore heterocomplexes H3/H4 and H2A/H2B (Figure 1B, upper panel). The histone pairing is done by three helixes of the fold domain of two histone counterparts which adopt a specific 'handshake' structure. The first high resolution crystal structure of the histone octamer in absence of DNA revealed that the histone octamer was organized in a tripartite structure wherein the H3/H4 complex formed a central tetramer which is flanked by two H2A/H2B dimers [6, 7](figure 1B, lower panel). Interestingly, while the histone fold domains were clearly resolved in the crystal, the unstructured tail domains were unseen. Although the histone octamer arrangement in presence of DNA confirmed the tripartite structure of the histone octamer, details of the edge of histone tails revealed the exit of these unstructured domains from the nucleosome [8].

It has been believed that the basic nature of the histones allowed the neutralization of the DNA phosphodiester backbone. However, the structure of the nucleosome at 1.9 Å resolution substantially improved the clarity of the electron density and revealed the presence of over 3000 water molecules and 18 ions [9]. The water molecules within the nucleosome promote the formation of hydrogen-bond bridges between the histone and the DNA molecule, like balls in a ball-bearing. Therefore, the water molecules enable the accommodation of intrinsic DNA conformational variation and promote the nucleosome mobility by limiting the rigidity of the nucleoprotein complex. The nucleosome crystal structures provided important information on the interactions between the histones and showed that the histone-DNA association is not only due to electrostatic interactions between the positively charge histones and the negatively charge DNA as it was primarily believed.

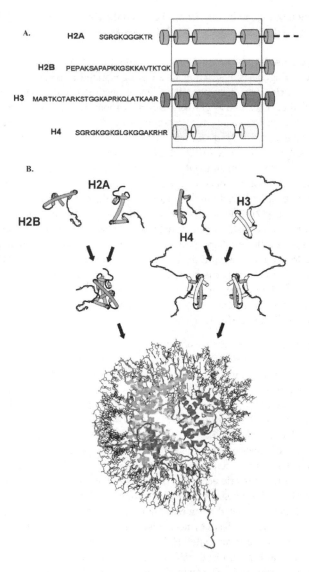

Figure 1. Histones and nucleosome formation: (A) schematic representation of the core histones. The boxes indicate the helixes of the histone fold domain, which is involved in the histone-histone interactions between H2A/H2B, and H3/H4. The amino-acid sequences correspond to the conserved sequence of the unstructured histone tail domain. (B) Individual core histones H2A (green), H2B (blue), H3 (yellow) and H4 (magenta) first heterodimerize to form the H2A/H2B and the H3/H4 complexes. The different complexes can either under different stringencies or with the help of histone chaperones associate together to form the nucleosome composed of a central tetramer of H3/H4 flanked by two heterodimers of H2A/H2B, and wrapped by 146 base pairs of DNA.

The demonstration of the labile interactions between the DNA molecule and the histone octamer was performed by the development of an elegant biochemical approach examining the accessibility of specific DNA sites within the nucleosomal DNA [10, 11]. In these experiments, the authors used a known nucleosome-positioning DNA sequence from the 5S gene, and by directed mutagenesis, restriction sites were generated at precise position within the DNA sequence. Nucleosome core particles were reconstituted with the different DNA sequences and purified by sucrose gradient centrifugation. The accessibility of the specific DNA sequences was examined as a function of time by adding to the nucleosome core particles the restriction enzymes. The quantitative analyses of the digested nucleosomal DNA reflect the accessibility of precise positions within the nucleosome core particles corresponding to the loss of histone-DNA contacts. Interestingly, the results revealed that DNA sequences engaged in the histone-DNA interactions are accessible to the restriction enzymes, and the accessibility gradually decreased when the restriction site is placed at proximity of the diad axis [12]. It was thus proposed that within the nucleosome core particle, dissociation of the histone-DNA contacts enables the transient exposure of DNA stretches to the solvent. Using a similar strategy, Widom and colleagues have also examined the contribution of the histone tail domains in the accessibility of nucleosomal DNA [13]. The results revealed that the removal of the histone tail domains leads to up to 14-fold increase in the site exposure within the nucleosomal DNA. Therefore, the tail domains within the nucleosome are also involved in the stabilization of DNA-histone fold domain interactions possibly by repressing the intrinsic dynamic nature of DNA.

The packaging of DNA in the nucleosome is a dynamic structure in conformational equilibrium, transiently exposing stretches of DNA off the histone surface, as demonstrated in model systems. Importantly, the binding of linker histone nearby the dyad axis to DNA restricts the flapping of the arms of DNA at the entry and at the exit of the nucleosome [14]. Although the analyses of the nucleosome behavior are very informative on the potential mobility of the nucleosome, it is obvious that the nucleosome is not recovered as a single subunit in living cell but rather found as a nucleosome polymer. Thus, the mobility of a considered monomer is possibly modulated by the surrounding nucleosomes. The analyses of a dinucleosome template generated from the 5S gene revealed a spontaneous mobility of the core histones which is restricted by the presence of the linker histone [15]. To better understand the function of the histones in the chromatin folding, it was required to examine templates that contained more than one or two nucleosomes. Using defined oligonucleosome models systems, the molecular mechanisms through which the histones modulated the chromatin folding were investigated [16]. These experiments revealed that the core histone tails play a critical function in the chromatin folding, as demonstrated by the removal of the tail domains *in vitro* [17, 18]. Interestingly, analyses of histone acetylation mimics on the chromatin fiber folding exhibited effects on the self-association properties of model nucleosome arrays, which depended upon the histone carrying the acetylation mimics and the number of mimics within the nucleosomes [19]. Such *in vitro* approaches using reconstituted nucleosomes systems are performed under particular pH and salt conditions. Additionally, acetylated histones increase chromatin solubility. Even if this can potentially biased the results, these investigations provide important features for understanding the physico-chemical parameters that facilitates or relieves the folding chromatin. But to date, the actual arrangement of the nucleosomes in the fiber is not yet well-determined. Nonetheless, experimental data have enabled to propose different models, the solenoid model and the zig-zag model, and it is possible that both models are juxtaposed in the nucleus [20, 21].

3. Relieving the chromatin inhibition

The ordered structure of chromatin represents the primary barrier to access the genetic information. On the basis of *in vitro* studies, the linker histones are proposed to be involved in the high-ordered chromatin structures [22]. Although the linker histone is not essential in protozoans [23, 24], the knock-out experiments in mouse revealed critical functions [25]. Indeed, in higher eukaryotes, the linker histones are composed of about eight subtypes which can compensate each other in some extend. However, upon the deletion of three subtypes, the synthesis compensation fails and embryonic lethality is observed. To attempt to gain insight in the function of the linker histone, analyses of the histone modifications have been carried out and reported a correlation between the cell cycle and the phosphorylation of the C-terminal tail domain [26, 27]. Surprisingly, while the genetic analyses revealed that preventing the phosphorylation of linker histone affects the chromatin organization leading to an increase of the nuclear volume, a raise in the linker histone phosphorylation was also detected in mitosis [28, 29]. Nonetheless, at the G1/S phase transition, linker histone is also found as substrate of cyclin-dependant kinase Cdk2, wherein the phosphorylation of the C-terminal tail leads to a relaxation of chromatin structure which might facilitate DNA replication [30, 31]. More recently, knock-down experiments of the linker histone in the slime mold *Physarumpolycephalum* showed a significantly faster rate of genome duplication, which was caused by a lost in the regulation of replication origin firing rather than the increase in the replication fork propagation [32]. Clearly, it has been evidenced that the linker histone affect the compaction of chromatin into the nucleus, and its release is required for the initial transition from non-permissive to permissive chromatin, but the actual mechanisms remain unclear.

Undoubtedly, if the primary inhibition for DNA replication is the higher levels of chromatin structure, relieving the high order of chromatin leaves the core histones associated with DNA, which is still an impediment for DNA accessibility. Thus, the next step is the release of the parental core histones to allow replication machinery to process all along the DNA molecule. To reach this goal, several concerns have to be taken into account. A bevy of studies have attempted to address the segregation of parental histones during replication, but the results are often controversial and many questions still need to be addressed. The fate of parental nucleosomes deals mainly with two overlapping key questions : do they dissociate from DNA during replication ? and, how are they transferred behind the replication fork ?

In vitro studies claimed chromatin replication without histone displacement. Initially showed in prokaryotic *in vitro* system [33], same conclusions were drawn from eukaryotic system studies [34]. In contrast, other studies evidenced that parental nucleosomes dissociate from DNA [35, 36]. The main argument for a non-displacement was that radioactively-labeled histone octamers are not reassembled onto a large excess of competitor DNA templates, suggesting that they do not dissociate from initial DNA matrix [34]. The idea that nucleosomes could partially relax to allow the passage of DNA processing machineries without complete dissociation is a matter of intense debate in the chromatin field, where the problematic of DNA accessibility is essential for most chromatin activities including replication, transcription and repair. Regarding replication, although no definite model can be drawn, it is commonly believed that disrupted parental nucleosomes are bound to specific protein chaperones which would transfer the core histone building blocks behind the replication fork.

The tripartite structure of the histone octamer implies that the removal of the H3/H4 from the nucleosome is associated with a displacement of the histone dimers H2A/H2B. However, two hypotheses could be postulated for lost of the nucleosomal structure, either the entire octamer is evicted or this is performed by the successive displacement of the different building blocks composing the histone octamer. Experimental approach for studying parental histone segregation implies the possibility of discriminating the old pool of histones and the new one [37]. By preventing the synthesis of new histones using translation inhibitors, like cycloheximide and puromycine, would enable the analysis of parental nucleosome transfer, though such treatments impair replication progression. Still, one can argue that as the replication process requires a tight regulation of the histone supply, impairing this regulation profoundly impact the replication leading to the replication fork blocks. Thus, most conclusions from these experiments have to be taken with caution. Original studies using this approach coupled with micrococcale digestion (enzyme allowing specific digestion of internucleosomal DNA) revealed that the size of the fragments obtained were consistent with DNA size protected by the histone octamer. So it was originally proposed that the parental nucleosomes are dissociated ahead of the replication fork and transferred behind with no detectable intermediate. Whether the experimental design led to artifacts remains likely.

Importantly, several studies using different approaches have demonstrated a distinct mobility for the H2A/H2B and the H3/H4 in living cells [36, 38]. On the basis of the different motions of the H2A/H2B and the H3/H4, one can reasonably believe that the octamer building blocks dissociate during cellular processes. Moreover, *in vitro* experiments for reconstituting or destabilizing nucleosome revealed the presence of basic heterocomplexes of H3/H4 tetramer and H2A/H2B dimer [37]. At physiological conditions, the heterotetramer H3/H4 prepared from chromatin and in absence of histone chaperones is the most stable form of the complex in solution [39]. Even if it has been claimed that a very transient dimeric state can exist, the absence of demonstration of the H3/H4 dimers led to the anchored view that parental nucleosomes split into two H2A/H2B dimers and a H3/H4 tetramer, and are then reassembled behind the fork, with the central tetramer H3/H4 deposited first [40, 41].

The simplest view regarding the dissociation of the parental core histone from DNA could be that the driving force of the replication fork progression is sufficient for overriding the histone-DNA interactions by the only action of replication specific proteins as helicases [42]. This model involves that core histones in presence of DNA spontaneously form nucleosomal structures with a tripartite organization. Unfortunately, *in vitro* experiments demonstrated that such arrangement of the histone octamer required either high salt concentrations or chaperone proteins to assist the proper loading of the histones in a tripartite structure [43]. A more comprehensive view was provided by a study by Groth *et al* [44] showing that the major H3/H4 histone chaperone ASF1 (Anti-Silencing Factor 1) forms a complex with the putative replicative helicase MCM2-7 (Minichromosome Maintenance Complex), via a H3/H4 bridge. On the basis of the *in vitro* capability of ASF1 to assemble chromatin, it has been proposed that this chaperone might be involved in the recycling and the transfer of parental H3/H4 histones directly coordinated by the DNA replication process.

Concerning H2A/H2B, picture is even less clear. Chaperones, like NAP1 (Nck-associated protein 1) and FACT (Facilitates Chromatin Transcription) might be involved. The heterodimeric complex FACT, a chromatin-modifying factor initially described to promote nucleosome rearrangement during RNA polymerase II-driven transcription through H2A/H2B dimer destabilization [45], was shown to be involved in DNA replication. FACT interacts with DNA polymerase α, and in human with the MCM helicase to act on DNA unwinding [46]. Recently, a conditional knock-out of one of the FACT subunit in DT40 chicken cells (Structure-Specific Recognition Protein 1, SSRP1) showed apparent impairment in replication fork progression [47]. Even if the precise mechanisms are still to be elucidated because this complex interacts with H2A/H2B and H3/H4 in multiple ways, the synergized action of histone chaperones and replication actors is actually an attractive model of coordinated nucleosome eviction/reassembly and DNA replication during S-phase.

It is known for a long time that chromatin assembly is an ATP-dependent process [48], so it is not surprising that ATP-dependent chromatin remodeling factors have been implicated in the release of the chromatin structure. Most studies focused on nucleosome movement during transcription, but strong arguments of their involvement during replication exist. The ISWI-class of ATP-dependent remodeling family interacts with several proteins in complexes, among them ACF1 (ATP-utilizing Chromatin assembly and remodeling Factor) and WSTF (Williams syndrome transcription factor). Depletion experiments demonstrated that ACF1 is critical for efficient DNA replication of highly condensed regions of mouse cells [49], and that WSTF, targeted to replication foci via its interaction with the processivity factor PCNA (Proliferating Cell Nuclear Antigen), promotes DNA replication by preventing premature maturation of chromatin [50].

4. Reforming chromatin behind the replication fork

Chromatin reassembly behind the replication fork is a rapid process. Electron microscopic studies and psoralen cross-linked nucleosome used, have clearly shown random distribution of the nucleosomal structures on both strand of the nascent DNA, with no apparent free-DNA [35]. By blocking protein synthesis with different inhibitors, it was demonstrated that half of the nucleosome pool came from random segregation of recycled parental ones, whereas the other half came from newly synthesized histones. In proliferating cells, the histone biosynthesis is coupled with the cell cycle progression. The vast majority of histones (the canonical histones) are massively produced at the beginning of the S phase, mainly by transcriptional activation of histone genes and improvement of pre-mRNA processing and stability, that begins during G1 phase (reviewed in [51, 52]). Through a feedback regulation reducing drastically the half-life of histone mRNAs, the amount of proteins then decreased at the end of S-phase until the baseline level is reached. However, experiments using replication blocking agents showed distinct synthesis profiles between H3/H4 and H2A/H2B, illustrating that specific level of regulation may exist [53]. Some specific histones (histone variants), used for deposition and exchange of nucleosomes outside of the S-phase (replication-independent chromatin assembly), are produced throughout the cell cycle. Although this aspect presents a great interest, the present chapter focuses on the regulation of the canonical histone proteins at the onset of DNA replication (for reviews about histone variants see [54, 55]).

Once the histones are synthesized, they are rapidly delivered to the site of replication and assembled into chromatin. Because these proteins are highly basic proteins, histones tend to promptly bind non-specifically to nucleic acids with a higher affinity to RNA than DNA, and they do not spontaneously form nucleosomes. To allow correct transfer into the nucleus and efficient deposition onto DNA, histone chaperones play a dual function, they neutralize the histone charge to prevent the formation of aggregates and they address the histones at precise locations within the nucleus [56].

The supply of histone is a tightly regulated process. Any events leading to replicational stress (as DNA damage for example) disturb the fine balance between histone supply and demand and have deleterious effects on the cell. Histone chaperone have critical roles in regulating this process. Consistently, deletion of the major histone H3/H4 chaperones CAF-1 (Chromatin Assembly Factor 1) or ASF1 in various organisms impair S-phase progression [57, 58]. In human, it was shown that ASF1 exists in a highly mobile soluble pool that buffered the histone excess [59]. In the budding yeast *S. cerevisiae*, ASF1 depletion impairs cell cycle progression and generates chromosome instability [60]. In this organism, it was shown that the up-regulation of the amount of histone in the cells leads to the degradation of the excess histones by a Rad53 kinase-dependent mechanism [61].

4.1. Transport into the nucleus

The nuclear import of the histone complexes is among the first levels of regulation. Several groups have attempted to define the mechanisms by which the histone supply might be regulated. The role of specific domains within newly synthesized histones essential for transport (and also formation of nascent chromatin) was first addressed using powerful genetic approaches in the yeast *S. cerevisiae*. Pioneer studies performed in budding yeast emphasized the essential role of both N-terminal H2A/H2B tails for cell viability (reviewed in [62]). Fusion protein experiments using fluorescent tracers led to the assumption that nuclear localization signals (NLS) are present in the N-terminal non-structured domain of histone proteins, and their interaction with karyopherin or importins would promote nuclear import of newly synthesized histones [63, 64]. Nevertheless, incorporation experiments of exogenous histones in the slime mold *Physarum polycephalum* showed that H2A/H2B dimers lacking both tail regions still localized to the nucleus. It was thus concluded that the tails of H2A/H2B are dispensable for nuclear import. However, the chromatin assembly analyses revealed that at least one tail is necessary for the deposition of the dimer complex into chromatin [65]. Conversely, studies using a similar strategy of incorporation of exogenous histones in *Physarum* to examine the fate of the H3/H4 complexes exhibited a function of the amino-terminal domains in nuclear import. Indeed, the histone H3/H4 dimers lacking H4 tail are inefficiently imported, while H3 tail was found dispensable in this process, but impaired nucleosome assembly coupled to replication [66].

By extending out of the nucleosomal structure, the exposed N-terminal regions of histones are subjected to active post-translational modifications. These marks, when imposed on assembled histones, have been shown to impact on the overall nature of the chromatin [67]. Newly synthesized histones are also characterized by a specific pattern of post-translational modifications, imposed in the cytoplasm shortly after synthesis. For example, newly synthesized H4 are diacetylated at lysine 5 and 12 by the holoenzyme HAT1 (Histone Acetyl Transferase 1), and these acetyl groups are rapidly removed after the assembly of histones into chromatin [68]. Despite the conservation of the H4 diacetylation throughout the evolution, the actual function in histone nuclear import and/or chromatin assembly remains undetermined. In Drosophila embryos, the RCAF complex comprises ASF1, acetylated H3K14, and diactetylated H4K5K12 [60] and in human, the CAC complex is composed of diacetylated H4K5K12 and CAF-1 [69]. This highlights an essential role of this dual signature for the formation of a complex between H3/H4 and the major chaperones associated to replication. However, as revealed by the co-crystal structure, ASF1 interacts with the C-terminal region of H3 [70], so the precise role of the post-translational modifications is not obvious. Strikingly, all described chaperones so far do not interact with the unstructured tails of histones. To conclude, even if the requirement of the amino-terminal regions of the histones has been evidenced for the assembly of chromatin and/or regulation of histones, their precise involvements in the overall process still necessitate investigations.

4.2. Mechanism of chromatin reassembly

Albeit the two DNA strands run in opposite directions, the progression of the replication fork is unidirectional. To reconcile that, during the replication process one daughter strand is synthesized continuously (the leading strand) whereas the other (the lagging strand) is build by short stretches of DNA named Okazaki fragments, ligated afterwards. Does this particular mode of duplication have an impact on parental nucleosomes segregation ? Even if adjacent "old" histones tend to segregate together, no clear preference for the leading or lagging strand have been demonstrated, mainly because the studies did not clearly discriminate the two strands. A recent study suggests that nucleosome positioning onto the lagging-strand could determined the length of Okazaki fragment in *S. cerevisiae*, via interaction with the enzyme polymerase pol δ, responsible for the extension of the nascent DNA chain through the 5' end of an Okazaki fragment [71]. By purifying Okazaki fragments, and performing the alignment onto the yeast genome, they demonstrated that they strikingly mapped with nucleosome positions. Once again, these experiments nicely illustrated the coupling between the DNA replication and the chromatin assembly.

The apparent higher sensitivity to nuclease digestion of newly synthesized chromatin compared to bulk chromatin suggests that new chromatin is not completely mature. Even though it was shown that specific post-translational modifications carried by newly synthesized histones and the absence of linker H1 histone could at least partially outline a more relaxed chromatin state, the reasons for the detection of the greater DNA accessibility in replicated chromatin remain actually elusive.

Newly synthesized H3/H4 are sequestered into the cytoplasm by ASF1, probably through interaction with several other chaperones, like the histone acetyltransferase HAT1, heat-shock proteins as HSC70 (Heat Shock Cognate 70 kDa protein), HSP90 (Heat Shock Protein 90), and NASP (Nuclear Autoantigenic Sperm Protein). The recent involvement of NASP as part of a cytosolic H3/H4 histone buffering complex is surprising, as this protein was initially described as an H1 chaperone [72, 73]. Indeed, in the nucleus ASF1 synergize with CAF1 via direct interaction with the p60 subunit. CAF1 was described to promote chromatin assembly *in vitro* [74]. This evolutionary conserved trimeric protein complex is recruited to site of DNA synthesis through interaction of the p150 subunit with the replication processivity factor PCNA, linking again chromatin assembly to replication fork progression [58]. As for parental histones, pioneer experiments using pulse-labeled histones suggested a sequential deposition of newly synthesized histones, with a H3/H4 tetramer assembled first, follow by the deposition of two H2A/H2B dimers.

The deposition model of nucleosomes, based on the stable tetrameric nature of histone H3/H4, was recently revisited [75]. Tagami and colleagues purified predeposition chromatin assembly complexes from HeLa cells stably expressing epitope-tagged histone H3.1 isoform (the replicative histone). The analyses of the immunoprecipitated tagged histones from purified nucleosomes and from the predeposition complexes showed that whereas about 50% of H3 in the nucleosomal fraction contained the epitope tag, all the histone complex in the predeposition complexes were tagged. It was thus concluded that H3/H4 complex is deposited onto DNA as dimer rather than tetramer. Biochemical, crystallographic and NMR analyses of ASF1 in complex with H3 (and sometimes H4) confirmed the dimeric nature of H3/H4 bound to the chaperone [70, 76, 77]. Furthermore, the structural data pointed out that the H3/H4 heterodimer binds ASF-1 at critical residues for H3/H3 interaction in the nucleosome, thus physically blocking the formation of a H3/H4 heterotetramer. This model has been reinforced by mutations of amino acids at critical regions. The dimeric nature of H3/H4 is also supported by a paper analyzing the composition of centromeric nucleosomes in the fruit fly *Drosophila*. At this particular genomic location, the nucleosome would exist in interphase as a stable tetramer, as a complex of single copy of CenH3-H2A-H2B and H4 has been identified [78].

5. Concluding remarks

The semi-conservative mode of replication of DNA ensures that the genetic information is faithfully transmitted to the daughter cells after mitosis. In higher eukaryotes, as the DNA is replicated, the chromatin environment has to be removed and subsequently restored. Here, we have reviewed an overview of the actual mechanisms that can sustain this operation. The studies described and cited in this chapter are based upon different experimental approaches, which might potentially present caveats inherent to the experimentations. Even though profound advancements have been reported during the past few years to clarify the factors involved in the transport and delivery of histones, basic concerns still have to be unraveled.

It is generally believed that the histone post-translational modifications impact chromatin structure and the chromatin activities through the recruitment of different effectors and modulators. Beside the mechanistic comprehension of the process of DNA replication in the chromatin context, underlying question addressed is how the chromatin organization and the information carried by histones are maintained or altered during replication. Indeed, the demonstration of the link between chromatin replication and cell differentiation suggests that the S-phase is a window of great opportunity for modulating the epigenetic regulations in a genetic program. However, in this context, the alterations of the chromatin structure and the histone modifications have not yet been fully elucidated.Three models can emphasize the nucleosome reorganization behind replication fork (Figure 2): (A) the entire parental octamer is transferred to form nucleosome and newly synthesized histones fill up the gaps. (B) The parental nucleosome splits into building blocks composed of a tetramer of H3/H4 and dimers of H2A/H2B, and the blocks are redistributed onto the two strands of DNA. The new histones are utilized for achieving the formation of the octamer. (C) The recently advanced dimeric nature of H3/H4 paved a new avenue for future investigations. The splitting of the tetramer could lead to mixed nucleosome, composed of parental and new histones.

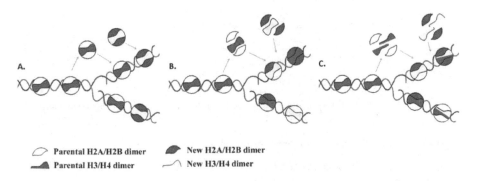

Figure 2. Working models of nucleosome reorganization during DNA replication. (A) Parental nucleosome is transferred as intact unit, without disruption of the octamer, leading to nucleosome fully constituted either of old or of new histones. (B) Parental nucleosome splits into H3/H4 tetramer and H2A/H2B dimers. In this model, new and old H3/H4 cannot coexist in the same nucleosome behind the replication fork. (C) Parental nucleosome splits into H3/H4 and H2A/H2B dimers, leading to mixed nucleosomes composed of old and new histones in each nucleosome building block.

In any considered model, the epigenetic information associated with the histone marks need to be copied from parental histones to newly synthesized ones. Concerning DNA methylation, the inheritance is a better-characterized process. In mammals, this modification mainly occurs on CpG dinucleotide (a cytosine followed by a guanine). The anti-parallelism of the DNA molecule, and the semi-conservative mode of DNA replication, ensure that the PCNA-interacting DNA methyltransferase DNMT1 easily copy the parental pattern onto the virgin daughter strand. To date, the mechanisms of the histone modification inheritance remains unclear. Most likely, future works in the field will attempt to address this issue.

Acknowledgements

The work in the Thiriet's lab is supported by grants from La Ligue contre le Cancer (44 ; 49 ; 53) and the national research agency (ANR).

Author details

Angélique Galvani[1,2,3] and Christophe Thiriet[1*]

*Address all correspondence to: Christophe.Thiriet@univ-nantes.fr

1 CNRS, UFIP (FRE 3478), Univ. of Nantes, Epigenetics: Proliferation and Differentiation, F-44322 Nantes, France

2 CNRS, UMR7216 Epigenetics and Cell Fate, F-75013 Paris, France

3 Univ Paris Diderot, Sorbonne Paris Cité, F-75013 Paris, France

References

[1] Mechali, M. (2001). Replicating. DNA in the Refractory Chromatin Environment. Nat Rev Genet, 2, 640-645.

[2] van Holde KE(1989). Chromatin. Springer-Verlag, New-York.

[3] Polo, S. E., & Almouzni, G. (2006). Replicating. DNA in the Refractory Chromatin Environment. Curr Opin Genet Dev, 16, 104-111.

[4] Felsenfeld, G. ((1978).) Chromatin. Nature ., 271, 115-122.

[5] Pruss, D., Hayes, J. J., & Wolffe, A. P. (1995). Nucleosomal anatomy--where are the histones? Bioessays, 17, 161-170.

[6] Arents, G., Burlingame, R. W., Wang, B. C., Love, W. E., & Moudrianakis, E. N. (1991). Replicating. DNA in the Refractory Chromatin Environment. Proc Natl Acad Sci U S A, 88, 10148-10152.

[7]   Arents, G., & Moudrianakis, E. N. (1993). Replicating. DNA in the Refractory Chro-
      matin Environment. Proc Natl Acad Sci U S A, 90, 10489-10493.

[8]   Luger, K., Mader, A. W., Richmond, R. K., Sargent, D. F., & Richmond, T. J. (1997).
      Replicating. DNA in the Refractory Chromatin Environment. Nature, 389, 251-260.

[9]   Davey, Sargent. D. F., Luger, K., Maeder, A. W., & Richmond, T. J. (2002). Replicat-
      ing. DNA in the Refractory Chromatin Environment. J Mol Biol, 319, 1097-1113.

[10] Polach, K. J., & Widom, J. (1999). Replicating. DNA in the Refractory Chromatin En-
      vironment. Methods Enzymol, 304, 278-298.

[11] Protacio, R. U., Polach, K. J., & Widom, J. (1997). Replicating. DNA in the Refractory
      Chromatin Environment. J Mol Biol, 274, 708-721.

[12] Polach, K. J., & Widom, J. (1995). Replicating. DNA in the Refractory Chromatin En-
      vironment. J Mol Biol, 254, 130-149.

[13] Polach, K. J., Lowary, P. T., & Widom, J. (2000). Replicating. DNA in the Refractory
      Chromatin Environment. J Mol Biol, 298, 211-223.

[14] Lee KM & Hayes JJ. (1998). Replicating. DNA in the Refractory Chromatin Environ-
      ment. *Biochemistry*, 37, 8622-8628.

[15] Ura, K., Hayes, J. J., & Wolffe, A. P. (1995). Replicating. DNA in the Refractory Chro-
      matin Environment. Embo J, 14, 3752-3765.

[16] Hansen, J. C., van Holde, K. E., & Lohr, D. (1991). Replicating. DNA in the Refractory
      Chromatin Environment. J Biol Chem, 266, 4276-4282.

[17] Fletcher TM & Hansen JC(1995). Replicating. DNA in the Refractory Chromatin En-
      vironment. J Biol Chem, 270, 25359-25362.

[18] Tse, C., & Hansen, J. C. (1997). Replicating. DNA in the Refractory Chromatin Envi-
      ronment. *Biochemistry*, 36, 11381-11388.

[19] Wang, X., & Hayes, J. J. (2008). Replicating. DNA in the Refractory Chromatin Envi-
      ronment. Mol Cell Biol, 28, 227-236.

[20] Woodcock, C. L., & Dimitrov, S. (2001). Replicating. DNA in the Refractory Chroma-
      tin Environment. Curr Opin Genet Dev, 11, 130-135.

[21] Schlick, T., Hayes, J., & Grigoryev, S. (2012). Replicating. DNA in the Refractory
      Chromatin Environment. J Biol Chem, 287, 5183-5191.

[22] Carruthers, L. M., Bednar, J., Woodcock, C. L., & Hansen, J. C. (1998). Replicating.
      DNA in the Refractory Chromatin Environment. *Biochemistry*, 37, 14776-14787.

[23] Shen, X., Yu, L., Weir, J. W., & Gorovsky, . (1995). Replicating. DNA in the Refractory
      Chromatin Environment. *Cell*, 82, 47-56.

[24] Shen, X., & Gorovsky, . (1996). Replicating. DNA in the Refractory Chromatin Envi-
      ronment. *Cell*, 86, 475-483.

[25] Fan, Y., Nikitina, T., Morin-Kensicki, E. M., Zhao, J., Magnuson, T. R., Woodcock, C.
      L., & Skoultchi, A. I. (2003). Replicating. DNA in the Refractory Chromatin Environ-
      ment. Mol Cell Biol, 23, 4559-4572.

[26] Dou, Y., Mizzen, Abrams. M., Allis, C. D., & Gorovsky, . (1999). Replicating. DNA in
      the Refractory Chromatin Environment. Mol Cell, 4, 641-647.

[27] Dou, Y., & Gorovsky, . (2002). Replicating. DNA in the Refractory Chromatin Envi-
      ronment. Proc Natl Acad Sci U S A, 99, 6142-6146.

[28] Bradbury, E. M., Inglis, R. J., Matthews, H. R., & Sarner, N. (1973). Replicating. DNA
      in the Refractory Chromatin Environment. Eur J Biochem, 33, 131-139.

[29] Bradbury EM, Inglis RJ, Matthews HR & Langan TA. (1974). Replicating. DNA in the
      Refractory Chromatin Environment. *Nature*, 249, 553-556.

[30] Alexandrow MG & Hamlin JL(2005). Replicating. DNA in the Refractory Chromatin
      Environment. J Cell Biol, 168, 875-886.

[31] Contreras, A., Hale, T. K., Stenoien, D. L., Rosen, J. M., Mancini, , & Herrera, R. E.
      (2003). Replicating. DNA in the Refractory Chromatin Environment. Mol Cell Biol,
      23, 8626-8636.

[32] Thiriet, C., & Hayes, J. J. (2009). Replicating. DNA in the Refractory Chromatin Envi-
      ronment. J Biol Chem, 284, 2823-2829.

[33] Bonne-Andrea, C., Wong, M. L., & Alberts, . (1990). Replicating. DNA in the Refrac-
      tory Chromatin Environment. *Nature*, 343, 719-726.

[34] Krude, T., & Knippers, R. (1991). Replicating. DNA in the Refractory Chromatin En-
      vironment. Mol Cell Biol, 11, 6257-6267.

[35] Sogo, J. M., Stahl, H., Koller, T., & Knippers, R. (1986). Replicating. DNA in the Re-
      fractory Chromatin Environment. The replication fork, core histone segregation and
      terminal structures. J Mol Biol, 189, 189-204.

[36] Jackson, V. (1990). Replicating. DNA in the Refractory Chromatin Environment. *Bio-
      chemistry*, 29, 719-731.

[37] Annunziato AT(2005). Split decision: what happens to nucleosomes during DNA
      replication? J Biol Chem, 280, 12065-12068.

[38] Kimura, H., & Cook, P. R. (2001). Replicating. DNA in the Refractory Chromatin En-
      vironment. J Cell Biol, 153, 1341-1353.

[39] Baxevanis AD, Godfrey JE & Moudrianakis EN(1991). Associative behavior of the
      histone (H3-H4)2 tetramer: dependence on ionic environment. Biochemistry, 30,
      8817-8823.

[40] Gruss, C., & Sogo, J. M. (1992). Chromatin replication. Bioessays, 14, 1-8.

[41] Gruss, C., Wu, J., Koller, T., & Sogo, J. M. (1993). Replicating. DNA in the Refractory Chromatin Environment. Embo J, 12, 4533-4545.

[42] Ramsperger, U., & Stahl, H. (1995). Replicating. DNA in the Refractory Chromatin Environment. Embo J, 14, 3215-3225.

[43] Wilhelm, F. X., Wilhelm, M. L., Erard, M., & Duane, M. P. (1978). Replicating. DNA in the Refractory Chromatin Environment. Nucleic Acids Res, 5, 505-521.

[44] Groth, A., Corpet, A., Cook, A. J., Roche, D., Bartek, J., Lukas, J., & Almouzni, G. (2007). Regulation of replication fork progression through histone supply and demand. Science, 318, 1928-1931.

[45] Belotserkovskaya, R., Oh, S., Bondarenko, V. A., Orphanides, G., Studitsky, V. M., & Reinberg, D. (2003). FACT facilitates transcription-dependent nucleosome alteration. Science, 301, 1090-1093.

[46] Tan, B. C., Chien, C. T., Hirose, S., & Lee, S. C. (2006). Functional cooperation between FACT and MCM helicase facilitates initiation of chromatin DNA replication. EMBO Jsj.emboj.7601271., 25, 3975-3985.

[47] Abe, T., Sugimura, K., Hosono, Y., Takami, Y., Akita, M., Yoshimura, A., Tada, S., Nakayama, T., Murofushi, H., Okumura, K., et al. (2011). The histone chaperone facilitates chromatin transcription (FACT) protein maintains normal replication fork rates. J Biol Chem, 286, 30504-30512.

[48] Glikin, G. C., Ruberti, I., & Worcel, A. (1984). Chromatin assembly in Xenopus oocytes: in vitro studies. Cell, 37, 33-41.

[49] Collins, N., Poot, R. A., Kukimoto, I., Garcia-Jimenez, C., Dellaire, G., & Varga-Weisz, P. D. (2002). An ACF1-ISWI chromatin-remodeling complex is required for DNA replication through heterochromatin. Nat Genet, 32, 627-632.

[50] Poot, R. A., Bozhenok, L., van den, Berg. D. L., Hawkes, N., & Varga-Weisz, P. D. (2005). Chromatin remodeling by WSTF-ISWI at the replication site: opening a window of opportunity for epigenetic inheritance? Cell Cycle, 4, 543-546.

[51] Marzluff WF & Duronio RJ(2002). Histone mRNA expression: multiple levels of cell cycle regulation and important developmental consequences. Curr Opin Cell Biol, 14, 692-699.

[52] Gunjan, A., Paik, J., & Verreault, A. (2005). Regulation of histone synthesis and nucleosome assembly. Biochimie, 87, 625-635.

[53] Loidl, P., & Grobner, P. (1987). Histone synthesis during the cell cycle of Physarum polycephalum. Synthesis of different histone species is not under a common regulatory control. J Biol Chem, 262, 10195-10199.

[54] Talbert, P. B., & Henikoff, S. (2010). Histone variants--ancient wrap artists of the epigenome. Nat Rev Mol Cell Biol, 11, 264-275.

[55] Hardy, S., & Robert, F. (2010). Random deposition of histone variants: A cellular mistake or a novel regulatory mechanism? Epigenetics, 5, 368-372.

[56] Hamiche, A., & Shuaib, M. (2012). Chaperoning the histone H3 family. Biochim Biophys Acta, 1819, 230-237.

[57] Mousson, F., Ochsenbein, F., & Mann, C. (2007). The histone chaperone Asf1 at the crossroads of chromatin and DNA checkpoint pathways. Chromosoma, 116, 79-93.

[58] Hoek, M., & Stillman, B. (2003). Chromatin assembly factor 1 is essential and couples chromatin assembly to DNA replication in vivo. Proc Natl Acad Sci U S A, 100, 12183-12188.

[59] Groth, A., Ray-Gallet, D., Quivy, J. P., Lukas, J., Bartek, J., & Almouzni, G. (2005). Human Asf1 regulates the flow of S phase histones during replicational stress. Mol Cell, 17, 301-311.

[60] Tyler, J. K., Adams, C. R., Chen, S. R., Kobayashi, R., Kamakaka, R. T., & Kadonaga, J. T. (1999). The RCAF complex mediates chromatin assembly during DNA replication and repair. Nature, 402, 555-560.

[61] Gunjan, A., & Verreault, A. (2003). A Rad53 kinase-dependent surveillance mechanism that regulates histone protein levels in S. cerevisiae. Cell, 115, 537-549.

[62] Ejlassi-Lassallette, A., & Thiriet, C. (2012). Replication-coupled chromatin assembly of newly synthesized histones: distinct functions for the histone tail domains (1) (1) This article is part of Special Issue entitled Asilomar Chromatin and has undergone the Journal's usual peer review process. Biochem Cell Biol.

[63] Mosammaparast, N., Jackson, K. R., Guo, Y., Brame, C. J., Shabanowitz, J., Hunt, D. F., & Pemberton, L. F. (2001). Nuclear import of histone H2A and H2B is mediated by a network of karyopherins. J Cell Biol, 153, 251-262.

[64] Mosammaparast, N., Guo, Y., Shabanowitz, J., Hunt, D. F., & Pemberton, L. F. (2002). Pathways mediating the nuclear import of histones H3 and H4 in yeast. J Biol Chem, 277, 862-868.

[65] Thiriet, C., & Hayes, J. J. (2001). A novel labeling technique reveals a function for histone H2A/H2B dimer tail domains in chromatin assembly in vivo. Genes Dev, 15, 2048-2053.

[66] Ejlassi-Lassallette, A., Mocquard, E., Arnaud, M. C., & Thiriet, C. (2011). H4 replication-dependent diacetylation and Hat1 promote S-phase chromatin assembly in vivo. Mol Biol Cell, 22, 245-255.

[67] Kouzarides, T. (2007). Chromatin modifications and their function. Cell, 128, 693-705.

[68] Annunziato AT & Hansen JC(2000). Role of histone acetylation in the assembly and modulation of chromatin structures. Gene Expr, 9, 37-61.

[69] Verreault, A., Kaufman, P. D., Kobayashi, R., & Stillman, B. (1996). Nucleosome assembly by a complex of CAF-1 and acetylated histones H3/H4. Cell, 87, 95-104.

[70] English CM, Adkins MW, Carson JJ, Churchill ME & Tyler JK(2006). Structural basis for the histone chaperone activity of Asf1. Cell, 127, 495-508.

[71] Smith, D. J., & Whitehouse, I. (2012). Intrinsic coupling of lagging-strand synthesis to chromatin assembly. Nature, 483, 434-438.

[72] Ransom, M., Dennehey, B. K., & Tyler, J. K. (2010). Chaperoning histones during DNA replication and repair. Cell, 140, 183-195.

[73] Hondele, M., & Ladurner, A. G. (2011). The chaperone-histone partnership: for the greater good of histone traffic and chromatin plasticity. Curr Opin Struct Biol, 21, 698-708.

[74] Smith, S., & Stillman, B. (1989). Purification and characterization of CAF-I, a human cell factor required for chromatin assembly during DNA replication in vitro. Cell, 58, 15-25.

[75] Tagami, H., Ray-Gallet, D., Almouzni, G., & Nakatani, Y. (2004). Histone H3.1 and H3.3 complexes mediate nucleosome assembly pathways dependent or independent of DNA synthesis. Cell, 116, 51-61.

[76] Agez, M., Chen, J., Guerois, R., van Heijenoort, C., Thuret, J. Y., Mann, C., & Ochsenbein, F. (2007). Structure of the histone chaperone ASF1 bound to the histone H3 C-terminal helix and functional insights. Structure, 15, 191-199.

[77] Natsume, R., Eitoku, M., Akai, Y., Sano, N., Horikoshi, M., & Senda, T. (2007). Structure and function of the histone chaperone CIA/ASF1 complexed with histones H3 and H4. Nature, 446, 338-341.

[78] Dalal, Y., Wang, H., Lindsay, S., & Henikoff, S. (2007). Tetrameric structure of centromeric nucleosomes in interphase Drosophila cells. PLoS Biol5, e218.

Telomeres

Chapter 8

Telomere Shortening Mechanisms

Andrey Grach

Additional information is available at the end of the chapter

1. Introduction

Telomeres are the terminal regions of the linear chromosomes of eukaryotes, which are composed of telomeric DNA and associated specific telomeric proteins. In most kinds of organisms, telomeric DNA is presented by a large number of repetitive, strictly defined short nucleotide sequences, such as: TTAGGG (in vertebrates), TTTAGGG (in the majority of terrestrial plants) and TTGGGG (in the ciliated infusoria Tetrahymena), etc. Although telomeric proteins differ among different groups of organisms they perform similar functions, which mainly consist of telomere length regulation and their protection against degradation (Grach, 2009). For a long time, it considered that telomeres did not code RNA molecules and thus proteins. Subsequently, it was found that RNA is still transcribed from telomeres but that it did not encode any proteins. Further studies showed that this RNA plays an important role in telomere length regulation and chromatin reorganisation during both development and cell differentiation (Azzalin et al., 2007). In spite of the fact that telomeres do not code proteins, they also perform very important functions, the main role of which is to maintain the stability and functionality of the cellular genome. Among these are the protection of chromosomes from fusion with each other (Blackburn, 2001), participation in mitotic and meiotic chromosome segregation (Conrad et al., 1997; Dynek and Smith, 2004; Kirk et al., 1997), the stabilisation of broken chromosome ends (Pennaneach et al., 2006), their attachment to the nuclear envelope (Hediger et al., 2002; Podgornaya et al., 2000), influencing gene expression (Baur et al., 2001; Pedram et al., 2006), counting the quantity of cell divisions (Allsopp et al., 1992; Kurenova & Mason, 1997; Olovnikov, 1973), and also an original buffer function (Olovnikov, 1973). The latter consists of the protection of the mRNA coding regions of chromosomes from the end replication problem. The end replication problem consists of the impossibility of the full reproduction of the previous length of linear DNA ends on the leading telomeres during of the S-phase of the cell division cycle. It is caused by peculiarities in their structure and the functioning of the DNA replication machinery. As a result the telomeric regions of chromosomes in daughter cells are shortened by several tens of nucleotides at each cell division (Lingner et al., 1995). In addition to the end replication problem, the telomere repair problem can also play a role in telomere length shortening. This problem in turn can be divided into the end repair problem and the shelterin-mediated telomere repair problem. The end repair problem includes the incomplete repair of DNA ends and direct damage-mediated telomere shortening and it can occur at the extreme ends of chromosomes. The incomplete repair of DNA ends consists in the inability of a repair system to complete the repair of damage if it occurred at the extreme ends of telomeres, since repair proteins cannot work correctly on the brink of a template and, as a result, they will also be shortened. Direct damage-mediated telomere shortening is based on the fact that, in some cases, the repair of damage at the extreme ends of chromosomes cannot even begin, in contrast to the incomplete repair of DNA ends at which the repair process begins but is not fully completed. It can generally be invoked by the fact that the breaks occurring on the extreme ends of chromosomes lead to the complete separation of the terminal DNA section and as a result repair system proteins are not able to even partially restore such damage. The consequence of this - and also of the subsequent actions of nucleases, which restore previous telomere ends' configuration - is the DNA ends shortening again. The shelterin-mediated telomere repair problem consists of the inability of the proteins involved in DNA damage response to detect and repair the damage occurring at telomeres due to the fact that the telomeric proteins in combination with telomeric DNA form a special structure on the telomeres called a telomeric loop (t-loop) that directly blocks DNA damage response proteins, as well as they block various DNA repair pathways themselves that is especially actual for the uncapped telomere condition when t-loop is not yet formed. This ultimately leads to the accumulation of damage and telomere shortening, and occurs along the entire length of telomeres where there is a telosome organisation and not just at their extreme ends as is the case with the end repair problem.

Telomere shortening is closely related to the replicative potential of cells and their lifespan. Thus, in accordance with A. M. Olovnikov's telomere theory of aging, when the telomere length approaches a certain critical level the cells stop dividing and begin ageing and are exposed to apoptosis upon reaching that level (Olovnikov, 1973). This fact has been confirmed experimentally in a number of studies (Allsopp et al., 1992; Allsopp et al., 1995; Aubert & Lansdorp, 2008). Besides playing a key role in aging, telomeres are also have great significance for carcinogenesis, as some cells with shortened telomeres acquire mechanisms to bypass the aging program and gain (among other characteristics) the ability to maintain telomere length and hence to "unlimited" quantity of divisions (Desmaze et al., 2003; Londoño-Vallejo, 2008; Stewart & Weinberg, 2006). The ability to elongate telomeres in vertebrates can be realised by means of two known mechanisms. The first and the most widespread mechanism among tumours provides for the use of a special enzyme called telomerase. It is a ribonucleoprotein enzyme consisting of a catalytic subunit, a telomerase RNA molecule and several additional components. Joining in with the ends of telomeres, its catalytic subunit uses reverse transcription of RNA, which is a part of telomerase to elongate a G-rich chain of telomeric DNA, which corresponds to the 3'-end regions of the DNA. Further, a C-rich chain corresponding to 5'-end DNA is synthesised on a template of a significantly elongated G-rich chain by a regular DNA polymerase reaction. As a result, the telomere ends gain the same structure as they had prior to the telomerase action but they become much longer in this case (Blackburn & Collins, 2011; Dong et al., 2005; Testorelli, 2003). The second mechanism, which is found in a minority of neoplasm types, is accomplished by recombination-mediated telomere replication and belongs to the alternative lengthening of telomeres mechanisms (ALT) (Grach, 2011a; Grach, 2011b; Henson et al., 2002; Muntoni & Reddel, 2005; Stewart, 2005). Besides the elongation of telomeres in tumour cells, telomerase also has a high activity in stem and germ cells, thereby providing them a high proliferative capacity (Meeker & Coffey, 1997). Meanwhile, its activity is low or absent in normal somatic cells, making their replicative capacity strictly limited (Rhyu, 1995). As for ALT, it is usually repressed in normal cells by telomeric proteins and certain other factors (Grach, 2011b).

In recent years, the study of telomeres has becomes increasingly popular among scientists who are engaged both with different branches of molecular biology as well as with the most distant problems in the whole of modern medicine. Such heightened interest in their study is first of all caused by the fact that telomeres perform very important functions in the maintenance of eukaryotic cell genome normal functionality. Besides this, telomeres define the replicative capacity of cells and play a key role in their aging and transformation processes which make these end structures an even more important subject for research. All of the above mentioned roles of telomeres depend upon their shortening and, therefore, telomere shortening mechanisms are among the key aspects of telomere biology, because the loss of chromosome functions, cell aging and degeneration are associated with the telomeric regions of chromosomes' length shortening. In this respect, the study of these mechanisms as well as the factors involved in their protection and elongation are of primary importance as long as our cumulative knowledge can help in the future in the struggle against aging, tumours and many other diseases, the treatment of which requires a high replicative capacity in the cells. Based on the great significance of the telomere shortening process, the nub of various telomere-shortening mechanisms will be considered in detail in this chapter, namely the end replication and telomere repair problems.

2. The end replication problem

2.1. Early views of the end replication problem

2.1.1. The end replication problem as a cause of telomere length shortening, which determines the replicative potential of cells

For the first time, the problem of eukaryotic linear chromosome ends' replication was proposed and described in detail in the form of the theory of marginotomy by A. M. Olovnikov in Russian in 1971 (Olovnikov, 1971). One year later, in 1972, the problem was also described by J. D. Watson, independently of Olovnikov (Watson, 1972). In 1973, the problem proposed by Olovnikov was represented in its English version (Olovnikov, 1973). At the heart of this problem, as was suggested, lies the inability of the usual DNA replication system to fully complete the replication of linear DNA ends in the process of cell division. As a result of this, it was assumed that the telomeric regions of chromosomes are shortened by roughly tens of base pairs (bps) at each cell doubling (Olovnikov, 1973). This state of affairs should explain why normal somatic cells, having divided a number of times, stop their further reproduction, start ageing and undergo apoptosis (i.e. the causes of the Hayflick cell division limit) (Hayflick, 1965). It was therefore suggested that in all of the non-transformed somatic cells of the organism, the telomere replication mechanism is absent and as a result of which they are gradually shortened on their division. When telomeres shorten to the definite minimal length needed for their normal functioning, the cells stop their division, age and then die. In other words, it was suggested that the telomere shortening process is a kind of "counter" which determines the replicative potential of cells (Olovnikov, 1973). These suppositions have been confirmed experimentally in several studies (Allsopp et al., 1992; Allsopp et al., 1995; Levy, et al., 1992). Thus, in one such experiment it was found that cells with shortened telomeres could perform far fewer divisions than cells with long telomeres (Allsopp et al., 1992).

2.1.2. The old theoretical model of the end replication problem

We now consider the actual mechanisms of the incomplete replication of the ends of linear DNA. As is known, every human chromosome consists of two anti-parallel DNA strands, which together form a single linear double-stranded DNA molecule with two ends. When the end replication problem was described for the first time, it was still considered that according to the generally accepted DNA model its ends would also have a double-stranded structure. Proceeding from this understanding of DNA, has been formulated the old theoretical model of the end replication problem, which was based on two possible mechanisms by means of which DNA ends could not be completely replicated.

The first mechanism suggests that DNA polymerase implementing DNA synthesis only in the $5' \rightarrow 3'$ direction should have besides the catalytic centre also the DNA binding site, which should be located in front of catalytic one and be responsible for attachment of enzyme to a parent DNA strand. As such, and during DNA replication, when a DNA polymerase approaches the very end of the template by moving in front of the DNA binding site it cannot continue synthesis and so dissociates from the DNA because it will have nothing more to bind to. As a result, the end portion of a template which is equal in length to a DNA binding centre cannot be replicated, since a DNA polymerase cannot simply bring its catalytic centre to the last nucleotides of a parent strand without being dissociated from the DNA. Thus, after an incomplete replication process of such a kind, the 3'-end of a new DNA strand should become shortened by several nucleotides when compared with the parental one (Olovnikov, 1973).

The second mechanism of the incomplete replication of DNA ends was based on the fact that a DNA polymerase is not able to begin new DNA synthesis itself but is capable only of elongating already existing oligonucleotides. Therefore, at the very beginning of replication, primase synthesises a short RNA primer of around 9-12 nucleotides long, which is subsequently elongated by a leading strand DNA polymerase. When the DNA polymerase has already synthesised a long enough DNA strand, the RNA primer is removed by RNase after that the gap is filled by the polymerase and the ends between the DNA fragments are connected by a ligase. The DNA end regions there do not form an exception and are replicated according to the same principle. However, the problem arises with the RNA primer, which is attached to the 3'-end of the DNA and defines the 5'-end of a new strand. The end gap arising after removal of the RNA primer cannot be filled later by the DNA polymerase, as there is no free 3'-end, which it could elongate. As a result, such incomplete end replication mechanism leads, this time, to the shortening of the 5'-end of a new DNA strand (Olovnikov, 1973).

Thus, according to the old theoretical model of the end replication problem, the 5' overhang at one end and the 3' overhang at the other can be formed at both daughter DNAs arising due to the peculiarities in the functioning of the DNA polymerase system. Such single-stranded protrusions should be cut by nucleases further in order to achieve double-stranded DNA ends' structure - which, as it was supposed earlier, they had initially. Consequently, the daughter chromosome telomere ends upon completion of the replication process will have the same configuration as they had before doubling. However, at the same time they will be shortened by a certain number of nucleotides (Olovnikov, 1973). The old theoretical model of the end replication problem is depicted in Figure 1.

2.1.3. Experimental confirmation of the old theoretical model of the end replication problem

As is shown in Fig. 1, the old theoretical model of the end replication problem assumes that incomplete replication can result from two mechanisms at both DNA ends - both on leading and lagging strands, resulting in each daughter chromosome being shortened at each of it's telomeres simultaneously. Further experimental verification of these circumstances has demonstrated quite different results for the leading strand and has completely confirmed the suppositions concerning the lagging DNA strand. In the course of one piece of research into the end replication problem using the artificially-created linear DNA replication system of the SV40 virus *in vitro*, it has been clearly determined that leading strand is synthesised entirely up to the very 5'-end of the template (Ohki et al., 2001). The explanation for this is that the DNA helicase unwinds its parent DNA to the very end and thus allows the completion of the synthesis of a new DNA strand. This discovery could abruptly undermine the possibility of the existence of the first proposed incomplete DNA replication mechanism, which assumes that due to the peculiarities of the functioning of DNA polymerase, the leading DNA strand synthesis cannot be fully completed up to the very end of the template. However, it is perhaps too early to judge this.

The results of another study of the end replication problem have shown the absolutely opposite situation. In the course of experiments looking into G-rich and C-rich single-stranded DNA in human fibroblasts, researchers unexpectedly revealed that the 5'-end of the DNA leading strand template is not replicated completely in the proliferating cells. Therefore, the 5'-overhangs appears in these cells predominantly during S-phase. This information provides grounds to suppose that the replication fork can terminate before reaching the chromosome's end. The authors of this study explain this in such a way that if the last RNA primer of the lagging strand is to be created as closely as possible to the 3'-end of the template then, in this case, the complete synthesis of the leading strand up to the very end of the 5'-end of a template is possible. If priming occurs more centromerically, then incomplete DNA leading strand replication and related to it, enhanced telomere shortening may be observed (Cimino-Reale et al., 2003).

Similar results have also been observed in one more study, although it was not carried out with nuclear DNA but rather with the linear mitochondrial DNA end regions of the yeasts (in humans, mitochondrial DNA is ring-shaped). It was found during this work that the DNA polymerase stops at a distance of approximately 110 nucleotides from the 5'-end of a template and does not continue further leading strand synthesis, thereby again leaving the 5'-overhangs. However, the authors of the research could not explain why this happens (Nosek et al., 1995).

Consequently, the results of the researches just reviewed are quite conflicting. Unlike the first investigation described, where DNA leading strand synthesis continues up to the very end of the 5'-end of the parent DNA, creating "blunt" end, in the second and in the third studies we observed the incomplete replication of the leading strand with the creation of a "sharp" end of the DNA molecule. Moreover, none of the studies describe the possibility that complete or incomplete replication of the parent DNA 5'-ends was due to the peculiarities in the function of DNA polymerase as had been suggested by Olovnikov in its first incomplete DNA ends replication mechanism. Instead they consider quite other reasons for - in one case of complete leading strand synthesis to the very end of a template, and in another instance of incomplete leading strand replication.

Therefore, this data cannot fully support or refute the possibility of the inability of DNA polymerase to complete the replication of the 5'-end through a failure to bring its catalytic site to the last nucleotides of a template. It is also very important to note that the above mentioned studies, which describe the incomplete synthesis of the DNA leading strand with the creation of a 5' overhang, in practice are almost unique in their kind. The prevailing majority of studies show that leading strand is synthesised completely up to the very end of a template (Chai et al., 2006; Lingner et al., 1995, Wright et al., 1997).

It was also shown in the experimental research reviewed by us initially that lagging strand synthesis stops within the area located at a distance of approximately 500 bps from the end of a parent strand leaving 3' overhangs there (Ohki et al., 2001). This in its turn fully supports assumptions concerning existence of the second mechanism of incomplete DNA replication described above. As too long extension of incomplete replication was found here, in this work the authors have reviewed somewhat in a new way this mechanism, performed on the lagging strand. It is known that the length of RNA primers range from 9 to 12 nucleotides, which has been described in most of the studies that we have analysed (Griep, 1995; Hao & Tan, 2002; Sfeir et al., 2005). In some cases, primers of 20-30 nucleotides in length are also mentioned (Bouche et al., 1978; Dai et al., 2009). Nonetheless, the length of an incomplete replication of a DNA lagging strand is much longer, and has been discovered to be as long as 500 nucleotides. Therefore, the authors of this research propose the following mechanisms for the incomplete replication of the DNA lagging strand. It should be noted that the first mechanism completely corresponds to that proposed in the old theoretical model of the end replication problem for DNA. As has already been noted, it is based on the removal of the end RNA primer and the further failure to fill the resulting gap with deoxyribonucleotides. As lagging strand's incomplete replication reached approximately 500 nucleotides and the RNA primer length ranges

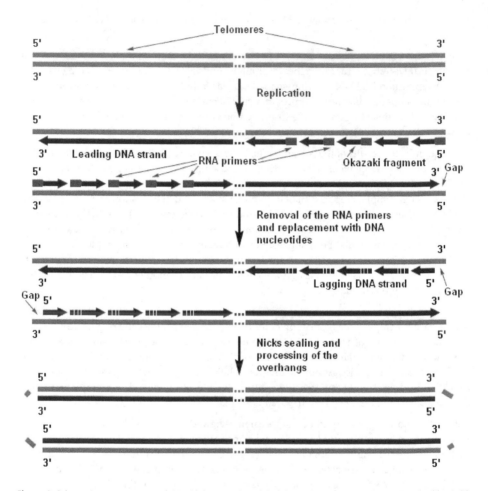

Figure 1. Schematic representation of the old theoretical model of the end replication problem, described by A. M. Olovnikov in 1971.

from 9 to 12 nucleotides, the authors of the work considered it very unlikely that this mechanism was the only one for the shortening of the 5'-end of daughter DNA. The second assumed mechanism consists of the inability of the DNA polymerase α-primase to begin lagging strand synthesis at the very end of a linear DNA molecule that can be the main cause of the end replication problem of the parent DNA's 3'-end (Ohki et al., 2001). It is also interesting that the length of Okazaki fragments, which represent the short DNA fragments with RNA primers at the 5'-end and are the key feature of the lagging strand, can range from between 100 to 500 nucleotides (Burgers, 2009; Mackenney et al., 1997; Zheng & Shen, 2011). Therefore, if such a fragment cannot be formed at the very end of the parent strand then the DNA daughter strand after the replication process will appear to be shortened in its length. Besides this research, many other studies showing similar results with incomplete lagging strand synthesis are also known.

Thus, most of the conditions of the old theoretical model of the end replication problem initially proposed by Olovnikov were confirmed absolutely in the course of the experiments carried out. In particular, it has been confirmed that telomeres are shortened at every cell division and that, specifically, these circumstances define the replicative potential of the cells and appear to be the cause of their aging and subsequent programmed cell death. Nevertheless, it was a failure to acquire any information absolutely confirming that the specifically incomplete replication of the DNA strands and subsequent cleavage of the resulting single-stranded overhangs of the molecule by nucleases results in telomeres' ends shortening, as was supposed by the old theoretical model. Unfortunately, we also failed to find any experimental data which fully confirms the first mechanism for the incomplete replication of DNA ends, assuming that DNA polymerase is not able to completely copy the 5'-end of the DNA leading strand template since it is incapable of bringing its catalytic site to the last nucleotides of a parent strand. At the same time, the results of the studies show that the 5'-end of a template remains in some cases not completely replicated, but other reasons for this, which are not directly related to DNA polymerase are specified in these cases. Therefore, the assumption regarding incomplete replication of the DNA leading strand is basically confirmed, but it is still unclear whether DNA polymerase directly plays a key role here or whether some other factors are involved (such as the ones that have been mentioned by the authors of the studies already discussed). Given all this, the assumption concerning the second mechanism of the incomplete replication of DNA ends is completely confirmed. Thus, the 3'-ends of the parent DNA, as was confirmed by the results of the experiments and initially stated in the theory, cannot be completely replicated during the lagging strand synthesis. As long as the extension of the incomplete replication of a lagging strand was much longer than the RNA primer length, it was supposed that the reason for the incomplete formation of a lagging strand along with the end primer removal might be due to the inability to prime and create the whole Okazaki fragment at the very DNA end. Given that the incomplete synthesis of the DNA leading strand is described only in some studies, but in the overwhelming majority of works it is shown that leading strand synthesis is performed completely up to the very end of a template, and that the 3'-end of the DNA template cannot for sure be replicated completely, the old theoretical model of end replication problem was also named a problem of incomplete lagging strand synthesis.

2.2. Modern views of the end replication problem

2.2.1. The establishment of the fact that telomere ends have a single-stranded structure

In the early 1980s, the data began to appear suggesting that both ends of each chromosome need not necessarily have a double-stranded structure but that they have a single-stranded structure (i.e. they are represented by 3'-overhangs). In 1981, it was noted for the first time that the ends of the linear minichromosomes, which are present in macronuclei of such ciliates as *Oxytricha*, *Stylonychia* and *Euplotes*, possess G-rich 3'-overhangs between 12 and 16 nucleotides long (Klobutcher et al., 1981). Later on, in 1989, similar results were also acquired for the linear extrachromosomal ribosomal DNA of ciliate *Tetrahymena* and - evolutionarily distant from it – the slim mould *Didymium* (Henderson & Blackburn, 1989). Later, in 1993, it was found that the telomeres of the yeast *Saccharomyces cerevisiae* also gain 3'-end overhangs in the late S phase of the cell cycle and which differ a little in their dimensions from the ones which were described in previous works, being formed by more than 30 nucleotides (Wellinger et al., 1993). Unlike the above-mentioned organisms, which have a constant G-overhang length, the telomeric overhangs of higher eukaryotes display variability, even among the different cells studied in one group. As has been demonstrated by the results of a great number of studies, human telomeres possess very heterogeneous 3' overhangs, ranging from very short ones 35 nucleotides long and even less, to very long ones with an extension of 500 nucleotides or more. Furthermore, such varying in their dimensions G-overhangs are observed in all types of examined cells including the telomerase-positive transformed cells, telomerase-negative normal mitotic cells and post-mitotic cells (Cimino-Reale et al., 2001; Makarov et al., 1997; McElligott & Wellinger, 1997; Stewart et al., 2003; Wright et al., 1997). All these observations allow the supposition that G-overhangs are a general feature of eukaryotic chromosome telomeres.

2.2.2. *A new theoretical model of the end replication problem*

Based on numerous experimental observations showing that telomere ends' structure is not double-stranded but single-stranded, J. Lingner et al. have shown that this situation considerably changes established views as to the end replication problem. In particular, they demonstrated that the second mechanism of incomplete DNA replication, based on last RNA primer removal, no longer necessarily appears to be a problem for DNA replication machinery and the cause of telomere shortening. As primer cutting all the same leads to the creation of a 3' overhang, which also existed prior to replication and which is a normal structural feature of chromosome ends, so no genetic informational loss occurs in this case. In this respect, the incomplete synthesis of the lagging strand up to the end of a template can be considered now to be the mechanism of normal single-stranded 3' overhang telomere ends' structural formation. At the same time, a problem arises with the leading strand synthesis. This is caused by the fact that in the course of replication on such telomeres, daughter chromosomes lose the 3' overhang which was present in the parent chromosome and in the absence of telomerase this will accordingly result in their shortening. Moreover, if it is not restored by this enzyme to its previous state then, and only in this case, in the next round of replication might be observed the problem of incomplete lagging strand synthesis and already associated to it DNA shortening (Lingner et al., 1995) (Fig. 2). Thus, the result of replication with the new theoretical model proposed by Lingner et al. is a formation of two daughter DNAs which have one "sharp" end with a 3' overhang forming due to lagging strand synthesis and the other "blunt" end (or a "sharp" one with a 5' overhang if we take into consideration the possibility of performing the first mechanism of the old model of the end replication problem proposed by Olovnikov) forming on the leading strand. In contrast to the earlier proposed theoretical model of the end replication problem where overhangs should be cut, now single-stranded 3'-end protrusions remain intact, forming the natural eukaryotic chromosomes ends. Given all this, if according to a new theoretical model leading strand synthesis results in the loss of 3' overhangs and the formation of "blunt" DNA ends, but the results of many experiments show that both chromosome ends have G-overhangs, and given that incomplete lagging strand replication assumes its formation only on one end, then there should also be a mechanism creating such an overhang on leading telomeres (Fig. 2).

Lingner et al. have also proposed two possible mechanisms for previous 3' overhang formation which also guarantee that DNA shortening, due to a problem of incomplete lagging strand synthesis, can never occur. The first mechanism presupposes that after the DNA replication process the end of the newly synthesised leading strand in the "blunt" DNA end can be elongated by telomerase and as a result a "sharp" DNA end, with a previous 3' overhang, will be restored. The possible caveat of this variant is that the DNA molecule with the "blunt" end acts here as a substrate for telomerase but telomerase is able to elongate only single-stranded ends rather than double-stranded ends as was found earlier *in vitro*. Nevertheless, the possibility that telomerase access could be provided in this case by helicases, nucleases or proteins binding single-stranded DNA has been considered. The second mechanism assumes that telomerase acts before the replication process, elongating the 3' overhang. It creates a template for the gap-filling synthesis of the complementary C-strand. As a result of the elongation of the 5'-end by conventional DNA replication machinery and the subsequent removal of the RNA primer, a telomere end region acquires a 3' overhang structure again but becomes much longer. Now, when the replication process approaches its end, the overhang on the leading strand is also lost but the genetic material no longer decreases, since before replication the parent strand of the leading strand was elongated. Further, the so-formed DNA "blunt" end and, in particular, its 5'-end region are exposed to processing by nucleases resulting in the formation of a "sharp" end with a previous 3' overhang that existed prior to the elongation by the telomerase (Lingner et al., 1995).

2.2.3. Experimental evidence for the new theoretical model of the end replication problem

These mechanisms describe the different possibilities for the creation of a previous 3' overhang and opposition to telomere shortening due to an inability during leading DNA strand synthesis to create such a single-stranded protrusion. However, both of these mechanisms are based on the possibility of telomerase action. At the same time telomerase activity is either very low in most normal human somatic cells or else is not observed at all (Rhyu, 1995). In spite of this, 3' overhangs are observed at both chromosome ends in normal cells (Makarov et al., 1997). There are also the results of focused experimental studies, showing that the removal of the genes coding for telomerase components does not affect the G-overhang structure considerably and this in turn also shows that the formation of such overhangs occurs irrespective of telomerase activity (Dionn & Wellinger, 1996; Hemann & Greider, 1999; Yuan et al., 1999). Besides, it was found during another study that these overhangs are exposed to cell cycle-regulated changes independent of telomerase activity (Dai et al., 2010). At the same time, telomerase in the cells where it is present is capable of elongating the 3' overhang after it is formed and thus make it like in the previous parental telomere. On this basis, it might be supposed in principle that previous 3' overhang in cells where there is no telomerase activity cannot be restored, but at the same time in its place a new overhang, by means of a special mechanism which will be reviewed later, is formed which results in the telomere shortening. It is should also be mentioned that if it were forever restored to a previous state with the telomerase participation that was assumed in the above described mechanisms, it would lead to telomeres not be shortened during the course of cell doubling. This is equivalent to the acquisition of the unlimited replication potential which is observed mainly in the transformed cells. At the same time, the second mechanism reviewed presupposes that after preliminary elongation by polymerases, a DNA "blunt" end formed during replication due to the impossibility of creating a 3' overhang through leading strand synthesis is exposed to treatment by nucleases which process its C-rich strand and thus create an overhang of a specific length. Recent studies suggest that such post-replication treatment of a parent strand by nucleases, independently of whether there is telomerase in the cells or not, seems likely to be the main mechanism of 3' overhang formation in the leading telomeres (Lenain et al., 2006; van Overbeek & de Lange, 2006; Wu et al., 2010). If the parent 3' overhang before replication were to be elongated by telomerase, then the nucleases activity would further lead to previous 3' overhang formation, i.e. telomere end length does not decrease upon that and even increases, and if not, then these enzymes will create a new 3' overhang resulting in the shortening of the telomere's length. It is important to note that there are studies, showing that telomerase elongates the 3' overhangs of the leading daughter telomeres (Chai et al., 2006). Therefore, the first of the above reviewed mechanisms can be considered more realistic for previous 3' overhang restoration, especially taking into account that the DNA's "blunt" end after replication is necessarily exposed to the nuclease's influence and only after this does it become accessible for telomerase. It is also important to note that if in a case of accomplishment of the first mechanism of the old model of the end replication problem a DNA " sharp" end with a 5' overhang will be formed, the telomere ends shortening in that case would be even greater, as the incomplete synthesis of the DNA leading strand up to the end of a template and - related to this - excessive post-replication processing will take place. The latter is caused by the situation that nucleases now, in order to create a 3' overhang, will not only remove a certain number of C-strand nucleotides as a part of the double-stranded DNA, but also its single-stranded overhang. The schematic representation of the new theoretical model of the end replication problem is presented in Fig. 2.

It is clearly shown in Fig. 2 that under the new theoretical model of the end replication problem the incomplete DNA lagging strand synthesis, as a consequence of the impossibility of creating an Okazaki fragment and the removal of an end RNA primer, no longer leads to the daughter telomere's shortening but appears instead to be a kind of mechanism of their normal 3' overhanging structures' renewal. At the same time, during the synthesis of the leading strand, the DNA replication machinery is not able to recreate such an overhang on other chromosome ends as for its synthesis there is simply no template. Therefore, the leading telomeres of daughter chromosomes with respect to a parent chromosome lose their 3' overhang, which can be the cause of their further shortening. Experimental evidence for the claim that it is 3' overhang loss, which really leads to telomere shortening is derived from one study where it was found that the length of this overhang completely agrees with the chromosome end regions' shortening rate (Huffman et al., 2000). Nevertheless, there is also the data from another study showing that the G-overhang length does not correlate with the telomere-shortening rate (Keys et al., 2004). The authors of the research suggest that besides the 3' overhang loss in the course of DNA replication; the telomere-shortening rate is also influenced by damage from oxygen free radicals (Keys et al., 2004). As a result of such replication, there occurs the

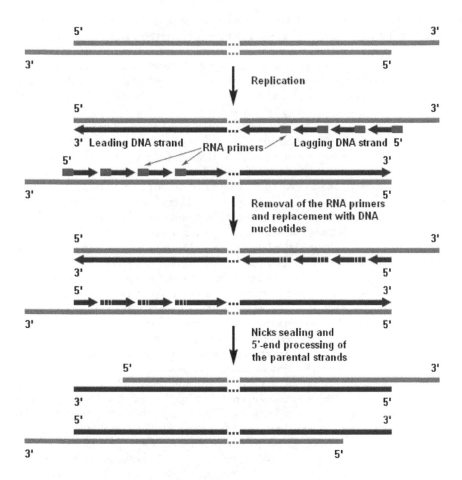

Figure 2. New theoretical model of the end replication problem.

formation of DNA daughter molecules that have one "sharp" end with a 3' overhang and the other "blunt" end. Taking into account that both chromosome ends have G-overhangs the "blunt" end on the leading telomere is further exposed to treatment by nucleases which cut its C-strand and thereby create the new 3' overhang, the length of which, will determine the rate of telomere shortening due to the end replication problem in the next replication cycle. The processing of the 5'-end of a parent strand of DNA can occur with the participation of such factors as the MRN protein complex (MRE11-RAD50-NBS1) as well as EXO1 and Apollo nucleases (Dewar & Lydall, 2010; Larrivee et al., 2004; Lenain et al., 2006; Maringele & Lydall, 2002; Tran et al., 2004; van Overbeek & de Lange, 2006; Wu et al., 2010; Zubko et al., 2004). Given this, the likely leading role is assigned to the Apollo nuclease, as RNA interference mediated repression of the gene encoding Apollo nuclease, leads to the loss of 3' overhangs, subsequent cell cycle arrest and programmed death (van Overbeek & de Lange, 2006). The dominant role of such overhangs on the ends of chromosomes, as has long been established in the course of experiments, consists of the formation of special structures called telomeric loops (t-loops, see below), which protect DNA ends from being recognised as double-stranded breaks by the repair system proteins and other enzymatic influences (Grach, 2009; Griffith et al., 1999; Stansel et al., 2001). This is why it is so important that the leading telomere form a new 3' overhang, even taking into account some DNA parent strand shortening. It should also be noted that if we assume - hypothetically - the possibility that chromosomes could function normally if a 3' overhang was present on only one DNA end, at the same time if in this case the parent 5'-end was never cut by nucleases, it would lead to the impossibility of telomere shortening in a considerable number of primary cells and their immortalisation without telomerase. Fortunately, it is not possible because then there would be a high probability of such cells transforming. As is shown in Fig. 2, telomere shortening as a result of one round of DNA replication happens due to the impossibility of recreating a 3' overhang on a leading daughter telomere and a subsequent undercut of a 5'-end by nucleases in creating such a new overhang. If, after these events, a telomerase does not become active and does not elongate this new 3' overhang, thereby providing the possibility of recreating a previous overhang, then the shortened 3'-ends, having passed to the next round of replication will lead to a situation such that now, on their template, even shorter 5'-ends will be created as a result of incomplete lagging strand synthesis. Therefore, it is possible to say that in such cases telomere shortening can be performed by means of incomplete lagging strand synthesis, which, it should be especially emphasised, is possible only after the previous loss of the 3' overhang or, as some have noted, after incomplete DNA leading strand synthesis, and so cannot proceed on the initial chromosomes. It is interesting to notice here that in certain scientific works, which describe the new model of the end replication problem, the 3' overhang loss is designated as a problem of incomplete leading strand synthesis. This description - as it appears to us - does not fully correspond to the actuality because, in this case, a template is replicated to the very 5'-end, unlike the variant proposed by the first mechanism of the old model of the end replication problem, where its incomplete replication with C-overhang formation can be carried out. Therefore, with regard to a single-stranded template, the leading strand synthesis here is performed completely, however the truth is that in relation to the parent double-stranded DNA with a 3' overhang on both sides it does not do so completely. Thus, irrespective of these descriptions, but in the case of absence of telomerase activity, it seems to be possible that the following chain of events under the new theoretical model of the end replication problem lead to a daughter telomeres' shortening: a 3' overhang loss on the DNA daughter strand, the processing of the 5'-end of the DNA parent strand within one round of replication, and then the incomplete replication of a shortened 3'-end of the previous DNA daughter strand in the next one.

In order to understand in more detail how incomplete DNA lagging strand synthesis and 3' overhang loss on the leading telomere is accomplished under the new theoretical model of the end replication problem, let us examine the structure of replication forks on both chromosome ends, as presented by Fig. 3.

Fig. 3 shows two replication bubbles on eukaryotic chromosome ends, each of which consists of a pair of replication forks moving in opposite directions. As is known, in most cases the initiation of replication is accomplished from a non-telomeric origin (Gilson & Geli, 2007). Later on, one fork of the replication bubble moves towards a centromere and another one towards a telomeric end. Here it is seen that helicase unwinds a double-stranded DNA molecule up to the very ends. It allows for DNA polymerases to finish leading strand synthesis completely, to the very end of a template. The double-stranded ends on the leading telomeres are formed upon that. As is clear from the figure, in such cases, the previous 3' overhang, which earlier was on the parent DNA ends, cannot be reproduced, in principle, in daughter molecules during leading strand formation as there is nothing for it to be synthesised on and, consequently, it is lost, resulting in telomere shortening. While the leading strand concerning the parent strand is synthesised completely, the lagging strand synthesis cannot be completed up to the end of the template. Earlier, it was thought that the removal of the end RNA primer is responsible for it. However, today many researchers are inclined to consider that this is not the only reason and it is also probably significantly complemented with the impossibility of creating the last Okazaki fragment. This situation is also well represented by Fig. 3. As is known, first of all the leading strand is synthesised in the motion of the unwinding of the parent DNA, and later on, after the DNA polymerase has synthesised a certain leading strand extension, it moves to a lagging strand and elongates it, thereby catching up with the first one. When such synthesis of both strands reaches the last point of unwinding - which can correspond to the 5'-end of the parent DNA - there remains a long stretch of single-stranded DNA in the form of a 3' overhang beyond its limits. Upon this, there is no more space for synthesising the leading strand in order that later on a DNA polymerase can move and fill such an overhang with a lagging strand. In this connection, the Okazaki fragment on the 3' overhang is not created and it remains non-replicated, and after the last primer removal its length increases a little more. However, as an overhang - which occurred prior to replication - is created anyway, the telomere's shortening does not happen in this case.

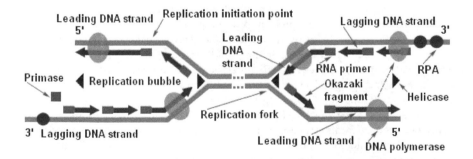

Figure 3. Telomeric replication forks.

Thereby, almost all conceptions of the new theoretical model of the end replication problem have been proved to be true in the course of the studies. It has been proved experimentally that 3' overhang loss in particular leads to telomere shortening. Besides it has also been confirmed that a new overhang is formed due to leading telomere processing by nucleases. At the same time, the assumptions as to the point at which a telomerase itself directly restores a previous overhang were not confirmed. For all this, according to new views of the end replication problem and which have been confirmed by a number of studies, from now on it should be construed not as the lagging strand synthesis problem but rather as the DNA leading strand synthesis problem.

2.3. The conclusion of this section

Thus, summarising all of the aforesaid, the views of the end replication problem as the cause of telomere shortening have changed over a period of several decades. Initially, when everyone considered that the structure of chromosome ends was double-stranded, it was supposed that telomeres were shortened mainly due to incomplete DNA lagging strand synthesis, which leads to the formation of 3' overhangs in support of which there is much experimental data. Furthermore, some studies have demonstrated the possibility of the accomplishment of incomplete leading strand synthesis up to the very end of a template with 5' overhangs forming, but the truth is that their number is small. Such overhangs further should be undercut, which would lead to the single-step shortening of chromosomes from both ends. As most of these works nevertheless provide evidence in favour of the idea that the replication of 5'-ends was carried out completely, it was later considered that telomeres were shortened only due to the problem of incomplete lagging strand synthesis. Here, it would seem that if the telomere shortening mechanism acts only from one end of a chromosome then the its other end would never decrease in length. Actually, this is not precisely true. The matter is that, if we were to monitor two strands of any initially parent DNA then one of them - after a certain number of divisions and in case of the absence of telomerase - will be really shortened from one end and the other one would be from the opposite end. If we continuously monitor some formed daughter strands, then in the subsequent generations of the cells there will also appear chromosomes which are shortened at their own two ends. After the establishment of the important circumstance that the structure of telomere ends is not double-stranded but 3' overhanging single-stranded, the problem of incomplete lagging strand synthesis already actually ceased to be treated as being the problem, since it no longer led to telomere shortening now, and only restored a previous configuration of their ends, which is important for normal chromosome functioning. At the same time, the existence of 3' overhangs on the chromosomes' ends creates a significant problem for leading strand synthesis. It is caused by the fact that in the course of replication, the 3' overhang which is present in the parent DNA on two ends cannot be renewed in the daughter DNAs during leading strand formation because of the absence of a template for its synthesis; in this connection it will be absent at one end in one daughter molecule and on another end of another one. Such a 3' overhang loss, the further processing of the 5'-ends of leading telomeres resulting in the formation of new G-overhangs and also the subsequent incomplete lagging strand synthesis in the next generation on a template of an already shortened 3'-end, actually lead to telomere shortening. In this connec-tion the end replication problem is inverted from the lagging strand to the leading strand. However, this does not mean that leading strand is necessarily synthesised incompletely up to the very end of a template, and it can be reproduced fully in this case. Thus, all of the observations described above have shaped our current thinking about telomere shortening during cell division.

3. The telomere repair problem

3.1. Early ideas of the telomere repair problem — The incomplete repair of double-stranded DNA ends

The problem of the incomplete repair of the very ends of DNA was also described, first, by A. M. Olovnikov as early as 1995. As with the old model of the end replication problem it was based on the idea that telomere ends have a double-stranded structure. The two suggested mechanisms of incomplete DNA ends' repair that are actually similar to the mechanisms of incomplete replication described in the previous section were distinguished. The essence of the first one concluded that if a single-stranded break (SSB) or "nick" occurred close to a 3'-end of a DNA strand at a distance of just several nucleotides, then this damage could not in principle be repaired. It was presumably connected with the following situation. The short end oligonucleotide created by the nick could not remain hybridized to the rest of the DNA molecule resulting in the formation of a gap with a protruding 5'-end. Later on, a repair DNA polymerase should attach to a DNA molecule and, while moving along the undamaged C-rich strand, synthesizes the lost 3'-end region on its template. However, it could not be performed in this case as DNA polymerase again, as well as with respect to the replication of the very 5'-end, not able to bring its catalytic site to the last nucleotides of a template in order to reproduce them on a complementary strand. Therefore, the DNA molecule remains shortened at the 3'-end afterwards. The second suggested mechanism provided for the impossibility of damage repair if a nick happened near the 5'-end of the DNA strand. A gap formed after the separation of a DNA fragment that was too short to remain hybridized to the template, could not be filled in for another reason in this case. The chain which was subject to repair and shortened due to the single-stranded break has no 3'-end or primer to which a DNA polymerase can add nucleotides in the course of repair synthesis and, therefore, should also remain non-elongated. On this basis, in both cases of the incomplete repair of DNA ends, single-stranded overhangs are formed which should be cut further by nucleases that would subsequently lead to telomere shortening (Olovnikov, 1995a; Olovnikov, 1995b; Olovnikov, 1995c).

3.2. The end repair problem — The incomplete repair of DNA ends and direct damage-mediated telomere shortening

The discovery that telomere ends had a single-stranded 3' overhanging structure, besides changing views on the end replication problem also considerably changed conceptions of the repair of chromosome ends. Before considering in detail exactly what these changes are characterised by, it is necessary to clearly define what should be understood by an incomplete DNA ends' repair. It is suggested by A. M. Olovnikov that it can proceed in two cases. In the first case, damage resulting in the breaking off of a single-stranded DNA fragment happens near the very end of the 3'-end of a double helix, and further DNA polymerase is not able to completely synthesise the lost region insofar as by moving along the opposite undamaged strand it cannot bring its catalytic site to the last nucleotides of the template. The second case assumes that damage with the separation of a DNA fragment happens near the 5'-end of a double-stranded molecule, and as a result DNA polymerase cannot restore the lost part again, since there is no available 3'-end as a primer to elongate. It is known that to begin polynucleotide synthesis at primer absence, as already noted, it is not able. Moreover there appears that the gap is so short that primase cannot even create an RNA primer. It is thus meant that in the first case a DNA polymerase binds to a DNA molecule and synthesizes several nucleotides, but the truth is that the last bases, which should lie opposite the very edge of the template, do not form (i.e. the damage repair process starts but does not come to its completion), and that in the second case the DNA polymerase, due to a RNA primer absence, cannot attach to a DNA double-stranded molecule and even to begin damage repair process. Based on these differences, specifically as to whether the repair process can start but does not come to its end or whether it does not start at all, we propose to look at the problem of incomplete DNA ends' repair and related to it telomere shortening somewhat in a new way. In particular, it is proposed that, under the problem of incomplete repair to consider that, the repair of damage at the end of double-stranded DNA starts but cannot finish completely because of the inability of most repair system proteins to correctly function at the very edge of a template that leads to chromosome ends shortening. It is necessary to emphasise that when noting that repair cannot be finished it is meant not only that DNA polymerase is unable to copy a template completely up to the very end in the course of repair DNA synthesis, but also that other enzymes can begin and even accomplish some stage of the repair process, however that later, due to certain reasons, repair cannot continue and so it finishes prematurely. As such, in the first case described, it is possible to say that what is actually accomplished is the incomplete repair of DNA ends. In addition, it is also proposed that if the repair of damage at a DNA end cannot start at all, again owing to the inability of repair proteins to work correctly at the edge of a template - and it will lead to chromosome telomere regions shortening, then this situation further should be designated as direct damage-mediated telomere shortening (DDMTS). The second described case can be related to this. Thus, we define two possible variants by means of which telomere shortening can be performed in a case where damage occurs at the very ends of a double-stranded DNA molecule, namely incomplete DNA ends' repair and direct damage-mediated telomere shortening.

Based on these new conceptions, let us consider once again Olovnikov's theoretical model of incomplete DNA ends' repair. As was mentioned, since it is described in the first variant that a DNA polymerase attaches to a DNA molecule, reproduces several nucleotides but subsequently cannot finish repair synthesis to the very end of a template, then in this case there occurs incomplete DNA repair. In the second variant, the enzyme cannot even attach itself to a template to begin repair - that was designated as DDMTS. It is necessary to note here that if, in the case of the first variant, the gap is very short then the DNA polymerase - even if it attaches itself to a template - will also not be able to begin repair because it will place its anchor region directly onto the very end of an undamaged DNA strand and, as a result, it will be immediately separated from the DNA molecule. This situation can already be regarded as DDMTS. At the same time, if with the second variant the gap will be long enough for an RNA primer to be formed, and then it is possible that there will be two variants, in both cases of which there will now occur the incomplete repair of the DNA ends. If the gap will be long enough to fit just an RNA primer, then in this case it might be supposed that when a DNA polymerase attaches itself, it will not be able to synthesise nucleotides as well, but as far as primase synthesizes the primer, then it is possible to consider that repair has started and that one of its stages has finished, but also that another one is not able to begin. In the future, such a primer is removed by RNase and a gap of the same length as it was before the repair arises. In the case where the length of a gap is such that in spite of the RNA primer several nucleotides are able to fit there, then the DNA polymerase synthesises them. However, after primer removal there will be a gap anyway but which the truth is that will be smaller than before the repair. As a result, such situation should also be viewed as incomplete repair. Ultimately, it is important to note that at replicative and cell senescence stages it is known that repair systems act poorly. Therefore, if a long enough gap appears at one of the DNA ends in senescent cells, then it will probably not be even partially repaired, and as a result DDMTS will take place. Thus, if such gaps are repaired in young cells, even if incompletely, then in old ones they will be not be repaired. It is also necessary to emphasise that at one time is apparently possible to separate only terminal single-stranded DNA fragment that was less than nine nucleotides in length, since it is widely known that RNA primers 9-12 nucleotides in size are strongly hybridized to a template. For that matter, when it was mentioned that a gap can arise is longer than primer itself or else the same, it can proceed only according to several steps, i.e. a successive separation of several fragments 8 nucleotides in size or less. Thus, if a break occurs, e.g. at a distance of 9 or more nucleotides, then such a terminal oligonucleotide will not only be able to hang on a template but will also be reunited with the remaining proximal part by DNA ligase. In summary, it is also necessary to add that since it was experimentally discovered that 5'-end copying by a DNA polymerase in the course of replication is, in most cases, accomplished completely, then in such a case only the second variant of incomplete repair based on RNA primer removal so described could feasibly be carried out.

3.3. Modern views of the end repair problem

Now let us consider what exactly are characterized by the changing of conceptions of repair at the ends of chromosomes, if the telomeres have single-stranded 3' overhanging structure. They are characterised by the following circumstances. First, given such telomere ends organisation, the problem of incomplete repair can arise, as it seems to be possible, only if a single-stranded break occurs at a distance of up to approximately eight nucleotides towards the centromere from a place where the 3' overhang begins and the 5'-end of complementary strand is situated (Fig. 4). In such a case, if a DNA polymerase even manages to copy a template up to the very 5'-end in the course of the repair synthesis of the lost single-stranded DNA fragment, the previous 3' overhang would still not be able to renew, and so it can be seen that in such a situation repair has begun but cannot be finished, insofar as the damage could not be fully repaired. If we take into account that a DNA polymerase might not be able to copy the last nucleotides of a template, then in such a case if a break with a subsequent separation of a DNA fragment occurs at a very short distance (e.g. of one nucleotide) from the above mentioned place, then repair will not begin and it will be designated as DDMTS. Secondly, if a break occurs somewhere at the 3' overhang or near its base (i.e. at a region where the opposite C-rich strand ends), then the distal part of the overhang or its entirely will separate from the DNA molecule and will be lost, as a result of which DDMTS will be observed - as far as repair in that case cannot even begin in principle due to single-stranded DNA fragment loss and the absence of a template for synthesizing the new one (Figure 5). With both variants, the new overhang will form in the future through 5'-end processing by the nucleases. Thirdly, an incomplete repair problem of the 5'-end, which should arise due to RNA primer removal on telomeres with "blunt" ends, is no longer a problem, and hence the reason for telomere shortening in instances with "sharp" ends, because as is the case with lagging strand synthesis at DNA replication, its cutting out leads only to the formation of the previous 3' overhanging configuration of telomere ends (Figure 6). Thus, single-stranded end breaks, at the 3' over-hanging telomere structure, can lead to incomplete telomere repair and further telomere shortening only if they will occur on a G-rich strand at a distance of several nucleotides in front of a place where a complementary C-strand ends. If the breaks affect the 3' overhang itself, then this will lead to DDMTS. Finally, the breaks of a C-rich strand occurring near the very 5'-ends will repair completely.

As is known, apart from single-stranded breaks, there are also such basic types of DNA damage as various nucleotide modifications, double-stranded breaks and cross-links (Sancar et al., 2004). Various nucleotide modifications (of a single one, a pair or else several) arising at DNA ends in most cases cannot lead to single-stranded breaks of the molecule in themselves. Further, they are exposed to various repair pathways, such as a direct repair (DR), a base excision repair (BER) and a nucleotide excision repair (NER) (Sancar et al., 2004). Since a direct repair is accomplished without any breakage of the phosphodiester backbone it cannot, in principle, lead to telomere shortening and, therefore, in our case, is of no particular interest. This type of reactions includes the photoreactivation of ultraviolet-induced pyrimidine dimers by a DNA photolyase enzyme, the removal of the O^6-methyl group from O^6-methylguanine (O^6MeGua) in DNA by the DNA methyltransferase enzyme, and the repair of apurinic/apyrimidinic sites through the direct insertion of bases by the insertase enzyme (Sancar et al., 2004). In addition, the repair of single-stranded DNA breaks by DNA ligase enzymes also belongs to this type but only if they do not arise at the very ends of DNA and do not lead to gap formation. Base excision repair consists of the cutting out of damaged nucleotide from a DNA strand by means of several reactions with the participation of DNA glycosylases, AP-endonuclease and phosphodiesterase, resulting in the formation of a very short gap (Fromme & Verdine, 2005; Krokan et al., 1997; Seeberg et al., 1995). This gap should be filled further by a DNA polymerase on a template of an undamaged complementary strand, after which the free ends are sealed by a ligase. If such a form of repair is carried out somewhere in the middle of the DNA molecule or near its 5'-end, then no problem will arise. However, if it proceeds at a distance of, say, 9 nucleotides from the place where the 3' overhang begins, i.e. the 9th nucleotide will be removed at that, then the end DNA fragment 8 nucleotides long up to the 5'-end of an opposite strand, together with the 3' overhang, would be lost. It will lead to gap formation, which can be filled further to form a "blunt" DNA end, but upon this, as well as in the case of a single-stranded break forming at a distance of up to 8 nucleotides and the subsequent formation of the same gap as discussed earlier, the previous 3' overhang it will not be able to restore itself and the telomere will shorten. Thus, base excision repair imposed on the end regions of a G-strand may be lead to an incomplete DNA repair. It is also interesting to discuss the situation where such a damaged nucleotide arises within the 3' overhang itself. BER system enzymes are apparently incapable of acting on a single-stranded DNA. As a result, such damage will not be repaired, and where further the base modification can similarly lead to DNA strand breaking and 3' overhang distal part loss, then this situation should be considered as DDMTS. However, if repair enzymes all the same could cut out a damaged nucleotide, which again will lead to the separation and loss of the 3' overhang terminal fragment, and so there will be an incomplete repair in this case. As for the nucleotide excision repair, it is very similar to BER but is accomplished by other enzymes, and in this case not only one damaged nucleotide is removed but up to thirty (de Laat et al., 1999; Reardon & Sancar, 2005; Sancar et al., 2004). With NER as well as with BER, if the gap is formed on the G-strand of telomeric DNA, in such a way that 8 or less nucleotides remain up to the end of a double-stranded DNA structure in the G-strand, then again a short end fragment of a G-strand together with the 3' overhang will be separated and lost and as a result an incomplete repair and related to it telomere shortening will be observed subsequently again. In order not to repeat this, it may be said that all other situations involving NER at the end of the telomere, including whether several nucleotides on the 3' overhang will be damaged, are similar to those that have been reviewed in relation to BER.

Double-stranded breaks (DSBs) of DNA can be repaired by three mechanisms: non-homologous end joining (NHEJ), microhomology-mediated end joining (MMEJ) and homology-directed repair (HDR) (Chu, 1997; Liang et al., 2008; Lieber et al., 2003). The first mechanism consists of the direct joining of broken ends by a specialised enzyme DNA ligase IV with the participation of the protein Ku and DNA-PK, which is carried out within G0/G1 and the early S phases of the cell cycle (Lieber et al., 2003). The second mechanism does not depend on these proteins and also differs from NHEJ in that this mechanism of DSBs repair uses 5-25 base pair microhomologous sequences to align the broken strands before joining, and it is carried out within the S phase of the cell cycle (Liang et al., 2008). The third mechanism is based on homologous recombination of a damaged chromosome with a sister chromatid or homologous undamaged chromosome, and therefore the damaged chromosome is repaired on their template that is carried out within the late S and G2 phases of the cell cycle (Chu, 1997). Nevertheless, NHEJ and HDR seem to be the main mechanisms for DSB's repair. If DSB occurs near the very end of a chromosome at a distance of only several nucleotides from the place where the 3' overhang begins, then such damage will probably not be repaired since the distal double-stranded fragment of DNA would be too short for the repair enzymes to bind to it, and, in the case of NHEJ and MMEJ, connect it to the rest of a molecule. Additionally, HDR also would be ineffective in repairing such damage since the homology tract would be again too short to effectively engage the enzymes that catalyze homologous recombination. As a result, the repair of such damage will not begin and there will be observed DDMTS. In addition, it is also necessary to note that if telomere damage occurs at a great enough distance from the very end of a DNA, but still within the telomere region organized by the telosome, it is unlikely that it could be repaired by homologous recombination. This is caused by the ability of telomeric proteins to block recombination events for the preventing of the elongation of chromosome telomere regions through ALT (Grach, 2011b).

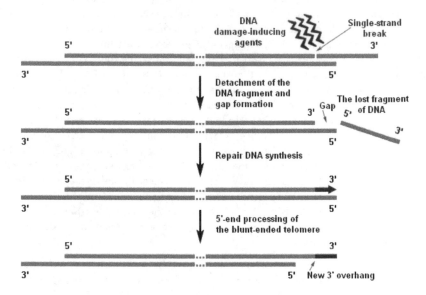

Figure 4. Incomplete double-stranded DNA end repair.

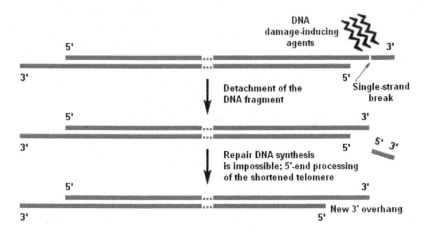

Figure 5. Direct damage-mediated telomere shortening.

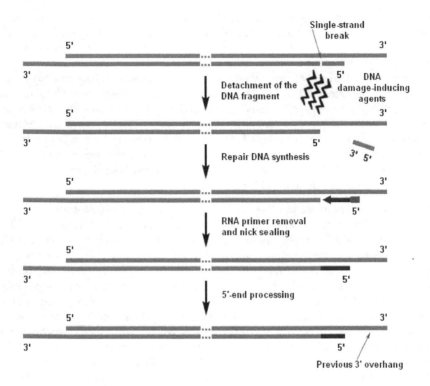

Figure 6. Complete double-stranded DNA end repair.

DNA cross-links represent connections between the nucleotide bases by covalent bonds (normally they are bound by hydrogen bonds) which can be of two types: intrastrand and interstrand (Dronkert & Kanaar, 2001). The latter ones represent a very serious problem for DNA replication machinery insofar as before a DNA polymerase replicates a parent DNA, it is at first unwound by helicase through the breaking of the hydrogen bonds between two strands. Therefore, if two complementary nucleotides are covalently linked, helicase will not be able to separate them and this will lead to the stalling of the replication fork and potential DSB formation. Two mechanisms for the repair of such damage are known which can be carried out during different phases of a cell cycle. Both of these mechanisms begin identically by forming cuts on both sides of the cross-link on one DNA strand by NER system proteins, after which there occurs a twisting of the oligonucleotide carrying the damage and gap formation. Later, such a gap will be filled; one mechanism for this is by TLS (Translesion synthesis), which makes use of DNA polymerases that are able to replicate DNA despite template damage arising before them. This is followed by another round of NER during which the second DNA strand is cleaved, and adduct removal is carried out. The second gap that is formed can than be filled by a conventional DNA polymerase on a complementary strand template and the ends are connected by DNA ligase. In the case of the other mechanism, the filling of the first gap is carried out during the course of recombination on a homologous chromosome template within a G2 phase of a cell cycle, upon completion of which NER proceeds again. The subsequent stages of repair are the same as those for the first mechanism. If cross-links occur somewhere on the non-telomeric chromosome regions, then these two repair mechanisms can act without any problems arising. If, however, they arise too close to an end of a double-stranded DNA, in such a way that there are 8 or less nucleotides left towards the place where the 3' overhang begins on a G-strand during gap formation, then NER system activity will lead subsequently to such overhang loss and incomplete repair with telomere shortening. It could probably be restored to a previous state only in the case of the recombination-mediated DNA synthesis, which on the very ends of chromosomes, is very tightly blocked, as has already been mentioned. Every other event here is similar to those that have been described in relation to BER and NER.

Thus, we have reviewed the possible cases of the occurrence of incomplete repair and of DDMTS for various types of the damage of chromosome ends with 3' overhangs, which in their turn, should lead to the telomere shortening. At the same, not all such possible cases (as well as variations of damage and mechanisms for their repair) have been analysed but only those that seem to be the most important. Also, it should be emphasized that every possible case of incomplete repair and DDMTS, which has been assumed, can arise on uncapped linear telomere ends. If telomere ends are in a capped condition (i.e. in the form of a t-loop) then already other such cases will probably be observed, which also will lead to their shortening. Nevertheless, it seems for us that if telomere ends are organised into t-loops then the cases of incomplete repair and DDMTS characterised by telomere shortening will occur much less frequently than with linear telomeres. It should be noted that experimental data fully and directly confirming the appearance of incomplete repair or DDMTS for different described variants of damage could not be found. At the same time, there are many studies providing general information, demonstrating that various kinds of damage can occur on telomeres, which are repaired much less efficiently than those which are formed on the non-telomeric chromosome regions, and that they lead to telomere shortening (Passos et al., 2007).

3.4. The shelterin-mediated telomere repair problem

Many investigations have been performed focussing on the influence of reactive oxygen species (ROS) on the occurrence of telomere damage and associated telomere shortening. In one of the earliest studies it was found that mild hyperoxia leads to accelerated telomere shortening and inhibits the proliferation of fibroblasts which, as it was supposed at that time, could happen due to the accumulation of single-stranded DNA breaks on chromosome end regions (von Zglinicki et al., 1995). In another study, it was found that oxidative stress really leads to the accumulation of single-stranded breaks on telomeres whereupon they actually become rapidly shortened (Petersen et al., 1998). In yet another study, it was found that ultraviolet radiation combined with riboflavin induces the formation of 8-oxo-7, 8-dihydro-2'-deoxyguanosine (8-oxodG) in DNA fragments with telomere sequences that further leads to the appearance of breaks in the area of the central guanine of GGG sequences. It was also shown that under the influence of hydrogen dioxide (H_2O_2) together with Cu (II) on these fragments DNA damage also occurred, which included the formation of 8-oxodG at the GGG sequence in the telomere sequence (5'-TTAGGG-3'), and which also led to breaks. Therefore, it was concluded that the formation of 8-oxodG in a GGG telomere sequence triplet induced by oxidative stress could play an important role in the acceleration of telomere shortening (Kawanishi & Oikawa, 2004). Along with these studies, many others showing that ROS leads to telomeric DNA damage formation and to their subsequent shortening are known (Passos & Von Zglinicki, 2006; Richter & von Zglinicki, 2007; Saretzki et al., 1999; Tchirkov & Lansdorp, 2003; Toussaint et al., 2000; von Zglinicki, 2000; von Zglinicki et al., 2000; von Zglinicki, 2002). It is also possible that besides the occurrence of single-stranded breaks on telomeres under the influence of ROS, there could also be oxidative modifications of nucleotides, which should be subject to repair by BER and NER systems, as well as double-stranded breaks (Passos et al., 2007). ROS has special importance concerning telomere damage, because unlike other mutagenic factors such as ionizing radiation, ultraviolet emanation, different chemical agents etc, ROS are constantly formed by mitochondria in a human organism during its normal metabolic activity. This is what triggered a strong interest to their study. Based on this, it may be that these other damaging agents can affect an organism and damage its telomeres in only very rare cases, while ROS continuously damages chromosomes' end regions, leading to their shortening. This situation actually should explain the fact that under normal conditions in the course of a cell's division, the telomere shortening rate considerably exceeds that which is expected only as a result of 3' overhang loss under the end replication problem (Keys et al., 2004). There are also studies which demonstrate that ROS can directly damage mitochondria themselves, and their mitochondrial DNA (mtDNA) in particular, thereby leading to their dysfunction which in turn can lead to the more intensive production of free radicals and, as a consequence, can result in even more intensive telomere damage and their shortening (Liu et al., 2002; Passos et al., 2006; Passos et al., 2007).

As was already noted, the damage occurring on telomeres is repaired less efficiently than that which originates in other genome regions (Kruk et al., 1995; Petersen et al., 1998; von Zglinicki, 2002). The reasons for such deficiencies in telomere-specific repair have not yet been completely established. At the same time, it is supposed that a basic role in the significant lessening of damage repair efficacy on telomeres belongs to the formation at their ends of the t-loops (capped telomere condition) (Passos et al., 2007). It was mentioned earlier that t-loops protect telomere ends from the activity of repair system proteins and another enzymatic influences (Grach, 2009; Griffith et al., 1999; Stansel et al., 2001). The example of the latter, incidentally, could be the telomerase attaching to the 3' overhang and its elongation. How does such repair suppression on telomeres by t-loops proceed? In order to answer this question let us first consider in detail what these t-loops represent and what actually characterises the response to DNA damage.

A t-loop represents a structure on eukaryotic chromosome ends which is formed at the bending back and subsequent insertion of a single-stranded telomeric DNA (3' overhang) into a double-stranded one (Grach, 2009). Upon this, the G-overhang forms a heteroduplex with the complementary C-strand region in double-stranded telomeric DNA, which is accompanied by the local untwisting of the latter and which leads to the formation of the so-called displacement loop (D-loop). The latter represents a triple-stranded structure which consists of a double-stranded DNA, formed by a 3' overhang and a C-strand of the previous double-stranded DNA, and single-stranded DNA, corresponding to the G-strand region of the previous double-stranded DNA (Fig. 7).

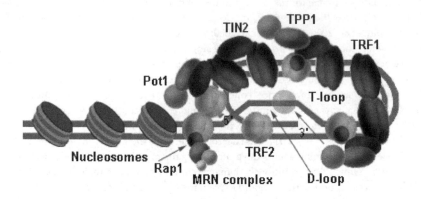

Figure 7. Structural organisation of a mammalian t-loop (Grach, 2009).

In t-loop formation, the primary role belongs to specific telomeric proteins, which are collectively referred to as the shelterin complex or telosome. Telomeric proteins differ slightly with different groups of organisms, but along with this they carry out similar functions. In mammals, shelterin includes six basic proteins, namely TRF1, TRF2, Rap1, TIN2, Pot1 and TPP1. These in turn can be divided into three groups: 1). double-stranded telomeric DNA binding proteins (TRF1 and TRF2); 2). single-stranded telomeric DNA binding proteins (Pot1); 3). proteins necessary for higher-order nucleoprotein complex formation (Rap1, TIN1 and TPP1) (Grach, 2009). The role of TRF1 function in t-loop formation it is to promote the bending back, twisting and linking of double-stranded telomeric DNA regions (Bianchi et al., 1997; Bianchi et al., 1999; Griffith et al., 1998; Griffith et al., 1999). Subsequently, TRF2 carries out the self-introduction of single-stranded telomeric DNA into a double-stranded one (Greider et al., 1999; Griffith et al., 1999; Stansel et al., 2001). Pot1 binds to single-stranded regions of telomeric DNA, which are represented only by a G-strand, and stabilises them (Baumann & Cech, 2001; Bunch et al., 2005; Churikov et al., 2006). Moreover, if a telomere end turns up in an uncapped condition, then Pot1 will cover the 3' overhang. If it turns up in a capped condition, then this protein will bind a single-stranded G-strand region in a D-loop structure. The Rap1 protein interacts with telomeres through binding with TRF2 (Li et al., 2000). It is suggested that the main functions of Rap1 are connected with its ability to recruit various repair system proteins, including Mre11/Rad50/Nbs1, Ku70/80 and PARP-1, to the telomeres (O'Connor et al., 2004). TIN2's role consists of both the binding of TRF1 proteins among themselves (Kim et al., 1999) and the binding of TRF1 with TRF2 (Ye et al., 2004). TPP1 is also necessary for the binding of TRF1 and TRF2 sub-complexes. It assists in the stabilising of TRF1-TIN2-TRF2 interaction. Besides this, TPP1 also directly binds Pot1 and regulates its activity (O'Connor et al., 2006). Thus, TIN2 and TPP1 play a key role in the association of different telomeric proteins in a single functional complex, which participates in the formation of t-loops and the capping of telomere ends, and this provides telomeres protection from different kinds of enzymatic action.

The DNA damage response is characterised by the following circumstances. It is possible to distinguish such basic key points as DNA damage detection from repair itself. Earlier, we considered the most important repair mechanisms, and therefore we will analyse the damage detection pathways. As is known, there are two distinct DNA damage detection pathways, which can potentially be activated by a chromosome's natural ends, namely the ATM kinase pathway and the ATR kinase pathway (de Lange, 2010). The ATM kinase pathway is triggered in response to double-stranded DNA breaks (Lee & Paull, 2007). Upon this, the process by which the ATM kinase response is accomplished is still not completely clear (Lee & Paull, 2007). It is known that the Mre11/Rad50/Nbs1 complex (sensor proteins which take part in double-stranded breaks' detection) binds to DNA ends upon this and activates the ATM kinase in a combination with Tip60 HAT (Carson et al., 2003; de Lange, 2010; Williams et al., 2010). Later on, the ATM kinase phosphorylates some key proteins involved in the damage response to DNA double-strand breaks and initiates the activation of the DNA damage checkpoint, which pauses the cell cycle and allows time for a cell to repair damage before continuing its division (Lee & Paull, 2007). Upon this, Tip60 HAT, through histone acetylation, modulates repair proteins' loading and repair of double-stranded DNA breaks (Murr et al., 2006). In that case, if the damage is not resolved, the p53 protein is then activated, which triggers an apoptosis program (Polyak et al., 1997). The ATR kinase pathway is activated in response to the single-stranded DNA, and is based on the point that the abundant single-strand DNA binding protein RPA recognises and associates with single-stranded DNA, resulting in an RPA-ssDNA complex (Cimprich & Cortez, 2008; de Lange, 2010; Nam & Cortez, 2011). Further, the ATR kinase together with the ATRIP protein recognises such a single-stranded DNA coated with RPA and attached to the DNA's damage site (Cimprich & Cortez, 2008). At the same, the ATR kinase's recruitment to the RPA-ssDNA complex strongly depends on the ATRIP protein, which itself directly attaches to RPA-ssDNA, and thus binds this complex with the ATR (Zou & Elledge, 2003). The checkpoint clamp complex containing RAD9-HUS1-RAD1 (9-1-1) proteins, which take part in checkpoint activation, cell cycle arrest, and recruitment of specific DNA polymerases and other repair proteins to damaged DNA is also independently recruited to the DNA damage site by RAD17 protein (checkpoint clamp loader) (Bermudez et al., 2003; Sohn & Cho, 2009). Besides, the ATR activator TOPB1 is recruited to the DNA damage site (Choi et al., 2009). After these events, the activation of ATR by TOPBP1 and the phosphorylation of downstream targets in a signal transduction cascade proceeds, which eventually leads to checkpoint activation, cell cycle arrest and subsequent damage repair (Cimprich & Cortez, 2008). Later on, all of the events are similar to those which were considered concerning the ATM kinase pathway, i.e. if damage is completely repaired then the cell cycle is resumed and the cell will continue its division, and if not then there will occur a trigger of the apoptosis program and subsequent cell death.

Now that we have considered what t-loops represent in themselves and what the DNA damage response is characterised by, it is possible to answer the question – how the repair of damage on telomeres is so strongly repressed. As such, it was proposed that shelterin hides the chromosome end from the ATM kinase pathway of DNA damage detection by remodelling telomeres into a closed structure – the t-loop. In a t-loop, Mre11/Rad50/Nbs1 is unlikely to recognise the telomere end as a double-stranded DNA end, which thus prevents ATM kinase activation, with subsequent cell cycle arrest and initiation of DNA damage repair (de Lange, 2009; de Lange, 2010; Griffith et al., 1999). This situation can be implemented as after the DNA replication, as well as after the double-stranded breaks occurred at telomeres. On the other hand, ATR signalling on telomeres is blocked by the shelterin Pot1 protein. It was noted earlier that, telomeres, on their own ends, contain a single-stranded DNA. This DNA at the uncapped condition of telomere ends is represented by 3' overhangs, and at the capped condition by a single-stranded region of a G-strand as a part of a D-loop. Furthermore, such single-stranded DNA can arise after a single-stranded damage and double-stranded breaks occurred at telomeres. It is potentially capable of activating the ATR kinase; however it was suggested that Pot1 binds a single-stranded telomere DNA and excludes the RPA protein from it. Later on, in the absence of this protein, such single-stranded DNA can no longer be distinguished by the ATR-ATRIP complex as damage, which prevents ATR kinase pathway activation and all subsequent events, including repair (de Lange, 2009; de Lange, 2010; Denchi & de Lange, 2007).

Along with the blocking of ATM and ATR kinase DNA damage detection pathways, shelterin can also block the DNA repair reactions by the direct blocking of repair proteins. It was mentioned earlier that the repair of double-stranded breaks can basically be performed by two pathways - NHEJ and HDR. These two pathways in turn are triggered in a manner similar to the ATM and ATR signalling pathways in various ways (de Lange, 2010). NHEJ first employs the ring-shaped Ku70/80 protein complex, which loads onto DNA ends and facilitates their further synapsis and ligation by DNA ligase IV. As such, there is the suggestion, that a t-loop in addition to the repression of the ATM signalling pathway, also - probably - effectively blocks Ku70/80 joining and thus could thwart NHEJ in its earliest steps (de Lange, 2009; de Lange, 2010; Palm & de Lange, 2008). Besides this, the possibility was discussed that additional mechanisms not involving the t-loop can be used for telomere ends' protection from NHEJ (de Lange, 2010). It is suggested that POT1 contributes in NHEJ repression, especially after DNA replication when the t-loop is not yet formed (de Lange, 2009). HDR is initiated when Rad51 (the protein playing one of the most important roles in homologous recombination, since it organizes the proteinaceous complex which is necessary for chromosomes pairing and subsequent DNA strands exchange (Babynin, 2007)) replaces RPA on a single-stranded DNA (de Lange, 2010). In this connection, for blocking such a pathway of DSB's repair at telomeres, it is enough simply to repress RPA binding. Such repression on the telomere ends of mammals is carried out with the help of the POT1 protein, the binding of which to a single-stranded telomeric DNA, as was already noted, excludes RPA (de Lange, 2010). Therefore, POT1 is probably capable not only of blocking the ATR signalling pathway, and NHEJ after DNA replication, but also HDR on telomeres (de Lange, 2010). Besides this, there is data suggesting that Ku70/80 is also capable of repressing HDR in the absence of POT1 (Celli et al., 2006). It is necessary to also note that there is experimental data showing that TRF2 overexpression weakens the repair of single-stranded breaks on telomeres, resulting in their accelerated shortening. This suggests the possibility that the repair of such damage on telomeres, as well as the other types of single-stranded damage, is again hindered by t-loops and shelterin, in the formation of which TRF2 participates. In this connection, at increased TRF2 production, the t-loops are probably formed more intensively in this case, and damages are repaired less effectively, leading to accelerated telomere shortening (Richter et al., 2007).

Thus, shelterin can inhibit repair on telomeres by the repression of various pathways of DNA damage detection, as well as of their repair itself. T-loop formation, in particular, leads to the blocking of the access of Mre11/Rad50/Nbs1 and Ku70/80 to double-stranded DNA ends, which prevents the activation of ATM signalling pathway and NHEJ that in its turn protects DNA natural ends, but blocks the repair of double-stranded breaks at telomeres. Besides, it is possible that the t-loop sterically blocks the repair of single-stranded damage at the telomeres by its three-dimensional structure. POT1 binding to single-stranded telomeric DNA excludes RAP from it and, therefore, prevents the recognition of damage by ATR in a complex with ATRIP, and which in turn prevents ATR kinase pathway activation. POT1 also blocks the binding of the Rad51 protein to single-stranded telomeric DNA, which prevents repair through HDR. Besides this, it is supposed that POT1 - after DNA replication when the t-loop yet is not formed - takes part in NHEJ repression, which is carried out, apparently, by Ku70/80 hetero-dimer blocking. There is also data suggesting that telomeric proteins themselves directly repress the pathways of single-stranded damage repair as well.

All of these mechanisms for repair repression on telomeres would seem, at first sight, to be the enemies for them, since repair deficiency leads to the circumstance where damage, occurring at chromosomes' ends is badly repaired and this leads to their shortening and, ultimately, leads to cell cycle arrest and apoptosis. Nevertheless, a certain amount of time is needed for this purpose, which in some cases can be a quite considerable. On the other hand, if telomeric proteins did not repress the DNA damage response at chromosomes' ends, it would result in apoptosis being triggered immediately rather than after telomeres had been shortened to a critical degree, which is caused by the following circumstances. If shelterin were be absent on distal telomere regions, or in other words if nucleosome organisation would be observed rather than telosome organisation, that probably would not distinguish them from other chromosome areas – it would lead to such a case whereby DNA natural ends would be recognised by the damage detection system as double-stranded breaks. In this connection, Mre11/Rad50/Nbs1 would activate subsequently the ATM kinase pathway. Besides this, it is known that MRN, attaching to double-stranded DNA ends and recognising them as DSBs, is also capable of performing the 5'-3' resection of such ends, thus creating 3' single-stranded tails or overhangs (Mimitou & Symington, 2009). With the absence of telosome and, in this case, of the POT1 protein in particular, this would now lead to the activation of the ATR kinase pathway. The activation of any of the DNA damage detection pathways would lead to cell cycle arrest and damage repair. The repair of such, let us say, false DSBs in the case of ATM signalling pathway would probably be carried out by the NHEJ way. Insofar as during NHEJ broken chromosome ends are directly joined, but in our hypothetical variant the natural ends of different chromo-somes would be recognized as broken ends, then NHEJ in this case could lead to various chromosomes' fusion with one another. This situation will result in genome instability, which leads to the initiation of the apoptosis program. The repair of damage by the ATR kinase pathway due to the presence of recombinogenic 3' single-stranded protrusions would possibly occur through HDR. In such a case, the fusion of different chromosomes and, later on, cell destruction through the initiation of the apoptosis program, will also take place. There is experimental data supporting all of this and showing that repression of the shelterin proteins leads to chromosomes' fusion and subsequent apoptosis. Besides this, such chromosome ends' vulnerability from homologous recombination could lead to continuous telomere elongation through the ALT which would essentially increase the probability of cell transformation. Thus, the repair suppression mechanisms on telomeres, although they lead to the accumulation of damage and shortening, at the same time protect chromosomes' ends from fusion among themselves and ALT activation. In the case of protection against chromosome fusion, it appreciably extends cells' lifespan. As for protection from ALT, thus shelterin prevents inappropriate telomere elongation and importantly the probability of cells' transformation. Therefore, the repair suppressing mechanisms on telomeres appear as friends for them.

It follows from the discussion above that the DNA damage response on telomeres, as well as damage detection and their repair pathways is potentially detrimental. This view was held for a long period of time. However, opinions changed when it was discovered that the proteins involved in the DNA damage response were present on the functional telomeres of mammals and interact with shelterin components, as previously in some way mentioned in reference to telomeric Rap1 protein, which recruits various proteins of the repair system to telomeres (Boulton & Jackson, 1996; Francia et al., 2006; Nugent et al., 1998). Moreover, further experiments have shown that they are also involved in telomere maintenance. This discovery was absolutely unexpected and somewhat paradoxical, as earlier it was thought that these proteins were the enemies for telomeres. However, it has now been discovered that this is not exactly the case, which has radically changed established views. Moreover, from now on DNA damage response proteins should be considered at the same time as equally the enemies and friends of telomeres. For the first time, such a role for DNA damage response proteins was found in yeast, where the NHEJ factor Ku is required for the maintenance of telomeres (Boulton & Jackson, 1996; Nugent et al., 1998; Polotnianka et al., 1998). In particular, in one such study it was found that in cells lacking telomerase but with functional Ku, telomere shortening slightly decelerates, i.e. it is less than in cells with repressed Ku function (Nugent et al., 1998). Further studies have shown that in addition to Ku, other proteins, involved in various DNA damage detection and their repair pathways, widely interact with telomeric proteins (Francia et al., 2006; Hsu et al., 1999; Hsu et al., 2000; Lenain et al., 2006; Palm & de Lange, 2008; Tarsounas et al., 2004; van Overbeek & de Lange, 2006; Zhu et al., 2000). Such factors in the mammals were called "shelterin accessory factors". Although they are present on telomeres transiently, at the same time they are very important for the maintenance of their normal structural organisation and functioning (Palm & de Lange, 2008).

Thus we emphasise, once again, that the main cause of poor damage repair on telomeres in comparison with other chromosome regions is the formation by their ends of t-loops, a process in which shelterin directly participates. Shelterin protects telomeres from the influence of both different sensor proteins that carry out DNA damage detection and the repair proteins themselves. Although this in turn leads to an accumulation of damage on telomeres and their shortening, at the same time it protects chromosomes from joining with each other, which ensures the maintenance of the integrity and normal efficiency of the cell genome. While shelterin protects telomeres from the action of DNA damage response proteins, it effectively interacts with them, which is also very important for telomere maintenance. It is necessary to add to all this also that in our opinion the cause of an inefficient damage repair on telomeres - specifically on their very ends - and their subsequent shortening, besides formation by telomeric proteins of the t-loops as well as their direct inhibiting influence on the DNA repair proteins is the fact that DNA repair proteins themselves cannot act correctly on the very DNA molecule end, which was discussed in the beginning of the section and can lead to incomplete repair or DDMTS.

3.5. The conclusion of this section

In summary, Olovnikov first described the problem of repairing the damage occurring on telomeres in the early 1990s. As at that time, it was considered that telomere ends had a double-stranded structure and so the model of telomere repair problem was described according to that conception. At that time, this problem was named the incomplete DNA repair problem, under which was considered the possibility of whether damage (DNA single-stranded break) occurred at a distance of only several nucleotides from the very end of a DNA molecule, then it could not be completely repaired in the course of copying of the complementary DNA strand because of DNA polymerases' functional peculiarities. As a result of this, telomeres should be shortened. In addition, based on that model, as well as on the point that in certain cases repair can start but does not come to completion or else does not begin at all, we considered it necessary to propose – an incomplete DNA repair and damage-mediated telomere shortening, or DDMTS. Upon this, incomplete repair and DDMTS - which in several different ways can lead to telomere shortening - can be considered to be two variants of the end repair problem. Considering these proposed new conceptions, the old model of the end repair problem was revised. As the time, it was found that the structure of telomere ends is not double-stranded, as was supposed earlier, but 3' overhanging single-stranded, we considered the eventualities of incomplete repair and the occurrence of DDMTS on the very linear telomere ends in instances of single-stranded breaks and other types of damage in accordance with this model. Theoretically, it is expected that incomplete repair and DDMTS for different variants of damage should uniquely lead to telomere shortening. However, it is not known whether it can actually occur, as unfortunately we were unable to find the experimental data confirming these assumptions. At the same time, there is much general data showing that damage on telomeres occurs and that this damage is repaired less effectively than damage in other regions of the chromosome. It was demonstrated that researchers' main attention in this has been given to telomere damage by reactive oxygen species, which are constantly formed in mitochondria at normal cell vital activity. It should explain why the genuine telomere-shortening rate exceeds the one that is expected as being only as a result of end replication problem. The circumstances by which the damage on telomeres is repaired more poorly has been explained well by many researchers in terms of t-loop formation. Therefore, in order to demonstrate exactly how t-loops repress repair, we considered what they represent in themselves and also what DNA damage response is characterised by. As a result, it was shown that t-loops, formed with the participation of the shelterin protein complex, as well as the telomeric proteins themselves, block various damage detection mechanisms and their repair pathways directly, and this actually causes telomere repair deficiency. This can be designated as the "shelterin-mediated telomere repair problem". It should be distinguished from the end repair problem, which can be carried out not along the whole telomere length but only on their ends, because of the inability of repair proteins to act on a template end. Thus, in principle, it is possible to identify two problems of repair on telomeres – the end repair problem, which is carried out near the very DNA molecule ends, and the shelterin-mediated telomere repair problem, which can affect all telomere regions where there is telosome organisation. At the same time, both of these problems can be referred to, in general, as the telomere repair problem.

4. Conclusion

In summarising all the data, it is necessary to emphasise that there exist two basic telomere-shortening mechanisms – the end replication problem and the telomere repair problem. The end replication problem, which is based on the 3' overhang loss during the course of DNA leading strand synthesis while the genetic material is doubling, has been studied in depth. The study of this problem allowed for the discovery of the telomerase enzyme and finding of the connection between telomere shortening and ageing, as well as carcinogenesis and various degenerative diseases. At the same time, many aspects of the end replication problem are still not absolutely clear. Therefore, further detailed studies of this process are necessary. The problem of telomere repair has been studied much less. Thus, there is no experimental data fully confirming that the end repair problem - which includes incomplete repair and DDMTS - can really be carried out on telomere ends and so lead to their shortening. At the same time, there is general data showing that damage on telomeres is frequently formed and repaired much less efficiently than on other chromosome regions leading to telomeric shortening. The researchers' main focus has been given to the influence of ROS on telomere damage. Given that they are constantly formed in cells by mitochondria, this is quite justifiable. At the same time, it is also necessary to study other negative factors concerning telomere damage and their shortening. Faint damage repair on telomeres is explained mainly by the formation at their ends of the t-loops, which are created with the participation of the telomeric protein complex shelterin and block different proteins involved in DNA damage response in order to prevent chromosomes' fusion with each other. This situation was referred to as the shelterin-mediated telomere repair problem. In this connection, it is necessary to study in more detail the structure of the proteins included in shelterin and their functions as well as mechanisms for t-loop formation. It is also important to study the proteins, which take part in DNA damage detection and the repair process itself. The latter should be studied in relation to both DNA damage response and their role in telomere maintenance. Thus, telomere-shortening mechanisms remain quite poorly understood and require further research.

Author details

Andrey Grach*

Khmelnitsky Regional Hospital, Ukraine

References

[1] Allsopp, R. C.; Vaziri, H., Patterson, C., Goldstein, S., Younglai, E. V., Futcher, A. B., Greider, C. W. & Harley, C. B. (1992). Telomere Length Predicts Replicative Capacity of Human Fibroblasts. *Proceedings of the National Academy of Sciences of the United States of America*, Vol.89, No.21, (November 1992), pp. 10114-10118.

[2] Allsopp, R. C.; Chang, E., Kashefi-Aazam, M., Rogaev, E. I., Piatyszek, M. A., Shay, J. W. & Harley C. B. (1995). Telomere Shortening is Associated with Cell Division *in vitro* and *in vivo*. *Experimental Cell Research*, Vol.220, No.1, (September 1995), pp. 194-200.

[3] Aubert, G. & Lansdorp, P. M. (2008). Telomeres and Aging. *Physiological Reviews*, Vol. 88, No.2, (April 2008), pp. 557-579.

[4] Azzalin, C. M.; Reichenbach, P., Khoriauli, L., Giulotto, E. & Lingner J. (2007). Telomeric Repeat Containing RNA and RNA Surveillance Factors at Mammalian Chromosome Ends. *Science*, Vol.318, No.5851, (November 2007), pp. 798-801.

[5] Babynin, E. V. (2007). Molecular Mechanism of Homologous Recombination in Meiosis: Origin and Biological Significance. *Tsitologiia*, Vol.49, No.3, (March 2007), pp. 182-193.

[6] Baumann, P. & Cech, T. R. (2001). Pot1, the Putative Telomere End-Binding Protein in Fission Yeast and Humans. *Science*, Vol.292, No.5519, (May 2001), pp. 1171-1175.

[7] Baur, J. A.; Zou, Y., Shay, J. W. & Wright, W. E. (2001). Telomere Position Effect in Human Cells. *Science*, Vol.292, No.5524, (June 2001), pp. 2075-2077.

[8] Bermudez, V. P.; Lindsey-Boltz, L. A., Cesare, A. J., Maniwa, Y., Griffith, J. D., Hurwitz, J. & Sancar, A. (2003). Loading of the Human 9-1-1 Checkpoint Complex onto DNA by the Checkpoint Clamp Loader hRad17-Replication Factor C Complex *in vitro*. *Proceedings of the National Academy of Sciences of the United States of America*, Vol. 100, No.4, (February 2003), pp. 1633-1638.

[9] Bianchi, A.; Smith, S., Chong, L., Elias, P. & de Lange, T. (1997). TRF1 is a Dimer and Bends Telomeric DNA. *The EMBO Journal*, Vol.16, No.7, (April 1997), pp. 1785-1794.

[10] Bianchi, A.; Stansel, R. M., Fairall, L., Griffith, J. D., Rhodes, D. & de Lange, T. (1999). TRF1 Binds a Bipartite Telomeric Site With Extreme Spatial Flexibility. *The EMBO Journal*, Vol.18, No.20, (October 1999), pp. 5735-5744.

[11] Blackburn, E. H. (2001). Switching and Signaling at the Telomere. *Cell*, Vol.106, No.6, (September 2001), pp. 661-673.

[12] Blackburn, E. H. & Collins, K. (2011). Telomerase: an RNP Enzyme Synthesizes DNA. *Cold Spring Harbor Perspectives in Biology*, Vol.3, No.5, (May 2011), pii: a003558. doi: 10.1101/cshperspect. a003558.

[13] Bouché, J. P.; Rowen, L. & Kornberg, A. (1978). The RNA Primer Synthesized by Primase to Initiate Phage G4 DNA Replication. *The Journal of Biological Chemistry*, Vol. 253, No.3, (February 1978), pp. 765-769.

[14] Boulton, S. J. & Jackson, S. P. (1996). Identification of a Saccharomyces cerevisiae Ku80 Homologue: Roles in DNA Double Strand Break Rejoining and in Telomeric Maintenance. *Nucleic Acids Research*, Vol.24, No.23, (December 1996), pp. 4639-4648.

[15] Bunch, J. T.; Bae, N. S., Leonardi, J. & Baumann, P. (2005). Distinct Requirements for Pot1 in Limiting Telomere Length and Maintaining Chromosome Stability. *Molecular and Cellular Biology*, Vol.25, No.13, (July 2005), pp. 5567-5578.

[16] Burgers, P. M. (2009). Polymerase Dynamics at the Eukaryotic DNA Replication Fork. *The Journal of Biological Chemistry*, Vol.284, No.7, (February 2009), pp. 4041-4045.

[17] Carson, C. T.; Schwartz, R. A., Stracker, T. H., Lilley, C. E., Lee, D. V. & Weitzman, M. D. (2003). The Mre11 Complex is required for ATM Activation and the G2/M Checkpoint. *The EMBO Journal*, Vol.22, No.24, (December 2003), pp. 6610-6620.

[18] Celli, G. B.; Denchi, E. L. & de Lange, T. (2006). Ku70 Stimulates Fusion of Dysfunctional Telomeres Yet Protects Chromosome Ends from Homologous Recombination. *Nature Cell Biology*, Vol.8, No.8, (August 2006), pp. 885-890.

[19] Chai, W.; Du, Q., Shay, J. W. & Wright, W. E. (2006). Human Telomeres Have Different Overhang Sizes at Leading Versus Lagging Strands. *Molecular Cell*, Vol.21, No.3, (February 2006), pp. 427-435.

[20] Choi, J. H.; Lindsey-Boltz, L. A. & Sancar, A. (2009). Cooperative Activation of the ATR Checkpoint Kinase by TopBP1 and Damaged DNA. *Nucleic Acids Research*, Vol. 37, No.5, (April 2009), pp. 1501-1509.

[21] Chu G. (1997). Double Strand Break Repair. *The Journal of Biological Chemistry*, Vol. 272, No.39, (September 1997), pp. 24097-24100.

[22] Churikov, D.; Wei, C. & Price, C. M. (2006). Vertebrate POT1 Restricts G-Overhang Length and Prevents Activation of a Telomeric DNA Damage Checkpoint but is Dispensable for Overhang Protection. *Molecular and Cellular Biology*, Vol.26, No.18, (September 2006), pp. 6971-6982.

[23] Cimino-Reale, G.; Pascale, E., Battiloro, E., Starace, G., Verna, R. & D'Ambrosio, E. (2001). The Length of Telomeric G-rich Strand 3'-Overhang Measured by Oligonucleotide Ligation Assay. *Nucleic Acids Research*, Vol.29, No.7, (April 2001), E35.

[24] Cimino-Reale, G.; Pascale, E., Alvino, E., Starace, G. & D'Ambrosio, E. (2003). Long Telomeric C-Rich 5'-Tails in Human Replicating Cells. *The Journal of Biological Chemistry*, Vol.278, No.4, (January 2003), pp. 2136-2140.

[25] Cimprich, K. A. & Cortez, D. (2008). ATR: an Essential Regulator of Genome Integrity. *Nature Reviews Molecular Cell Biology*, Vol.9, No.8, (August 2008), pp. 616-627.

[26] Conrad, M. N.; Dominguez, A. M. & Dresser, M. E. (1997). Ndj1p, a Meiotic Telomere Protein Required for Normal Chromosome Synapsis and Segregation in Yeast. *Science*, Vol.276, No.5316, (May 1997), pp. 1252-1255.

[27] Dai, H.; Liu, J., Malkas, L. H. & Hickey, R. J. (2009). Characterization of RNA Primers Synthesized by the Human Breast Cancer Cell DNA Synthesome. *Journal of Cellular Biochemistry*, Vol.106, No.5, (April 2009), pp. 798-811.

[28] Dai, X.; Huang, C., Bhusari, A., Sampathi, S., Schubert, K. & Chai, W. (2010). Molecular Steps of G-Overhang Generation at Human Telomeres and its Function in Chromosome End Protection. *The EMBO Journal*, Vol.29, No.16, (August 2010), pp. 2788-2801.

[29] de Laat, L.; Jaspers, N. G. & Hoeijmakers, J. H. (1999). Molecular Mechanism of Nucleotide Excision Repair. *Genes and Development*, Vol.13, No.7, (April 1999), pp. 768-785.

[30] de Lange, T. (2009). How Telomeres solve the End-Protection Problem. *Science*, Vol. 326, No.5955, (November 2009), pp. 948-952.

[31] de Lange, T. (2010). Telomere Biology and DNA Repair: Enemies with Benefits. *FEBS Letters*, Vol.584, No.17, (September 2010), pp. 3673-3674.

[32] Denchi, E. L. & de Lange, T. (2007). Protection of Telomeres through Independent Control of ATM and ATR by TRF2 and POT1. *Nature*, Vol.448, No.7157, (August 2007), pp. 1068-1071.

[33] Desmaze, C.; Soria, J. C., Freulet-Marrière, M. A., Mathieu, N. & Sabatier, L. (2003). Telomere-Driven Genomic Instability in Cancer Cells. *Cancer Letters*, Vol.194, No.2, (May 2003), pp. 173-182.

[34] Dewar, J. M. & Lydall, D. (2010). Telomere Replication: Mre11 Leads the Way. *Molecular Cell*, Vol.38, No.6, (June 2010), pp. 777-779.

[35] Dionne, I. & Wellinger, R. J. (1996). Cell Cycle-Regulated Generation of Single-Stranded G-Rich DNA in the Absence of Telomerase. *Proceedings of the National Academy of Sciences of the United States of America*, Vol.93, No.24, (November 1996), pp. 13902-13907.

[36] Dong, C. K.; Masutomi, K. & Hahn, W. C. (2005). Telomerase: Regulation, Function and Transformation. *Critical Reviews in Oncology Hematology*, Vol.54, No.2, (May 2005), pp. 85-93.

[37] Dronkert, M. L. & Kanaar, R. (2001). Repair of DNA Interstrand Cross-Links. *Mutation Research*, Vol.486, No.4, (September 2001), pp. 217-247.

[38] Dynek, J. N. & Smith, S. (2004). Resolution of Sister Telomere Association is Required for Progression Through Mitosis. *Science*, Vol.304, No.5667, (April 2004), pp. 97-100.

[39] Francia, S.; Weiss, R. S., Hande, M. P., Freire, R. & d'Adda di Fagagna, F. (2006). Telomere and Telomerase Modulation by the Mammalian Rad9/Rad1/Hus1 DNA-Damage-Checkpoint Complex. *Current Biology*, Vol.16, No.15, (August 2006), pp. 1551-1558.

[40] Fromme, J. C. & Verdine, G. L. (2004). Base Excision Repair. *Advances in Protein Chemistry*, Vol.69, (n.d.) pp. 1-41.

[41] Gilson, E. & Geli, V. (2007). How Telomeres are Replicated. *Nature Reviews Molecular Cell Biology*, Vol.8, No.10, (October 2007), pp. 825-838.

[42] Grach, A. A. (2009). Structural Organization of Telomeres in Various Kinds of Organisms. *Tsitologiia*, Vol.51, No.11, (November 2009), pp. 869-879.

[43] Grach, A. A. (2011a). Alternative Telomere-lengthening Mechanisms. *Cytology and Genetics*, Vol.45, No.2, (March-April 2011), pp. 121-130.

[44] Grach, A. A. (2011b). The Role of Alternative Lengthening of Telomeres Mechanisms in Carcinogenesis and Prospects for Using of Anti-Telomerase Drugs in Malignant Tumors Treatment. *Tsitologiia*, Vol.53, No.10, (October 2011), pp. 3-15.

[45] Greider, C. W. (1999). Telomeres do D-Loop-T-Loop. *Cell*, Vol.97, No.4, (May 1999), pp. 419-422.

[46] Griep, M. A. (1995). Primase Structure and Function. *Indian Journal of Biochemistry and Biophysics*, Vol.32, No.4, (August 1995), pp. 171-178.

[47] Griffith, J.; Bianchi, A. & de Lange, T. (1998). TRF1 Promotes Parallel Pairing of Telomeric Tracts *in vitro*. *Journal of Molecular Biology*, Vol.278, No.1, (April 1998), pp. 79-88.

[48] Griffith, J. D.; Comeau, L., Rosenfield, S., Stansel, R. M., Bianchi, A., Moss, H. & de Lange, T. (1999). Mammalian Telomeres End in a Large Duplex Loop. *Cell*, Vol.97, No.4, (May 1999), pp. 503-514.

[49] Hao, Y. H. & Tan, Z (2002). The Generation of Long Telomere Overhangs in Human Cells: a Model and its Implication. *Bioinformatics*, Vol.18, No.5, (May 2002), pp. 666-671.

[50] Hayflick, L. (1965). The Limited *in vitro* Lifetime of Human Diploid Cell Strains. *Experimental Cell Research*, Vol.37, (March 1965), pp. 614-636.

[51] Hediger, F.; Neumann, F. R., Van Houwe, G., Dubrana, K. & Gasser, S. M. (2002). Live Imaging of Telomeres: yKu and Sir Proteins Define Redundant Telomere-Anchoring Pathways in Yeast. *Current Biology*, Vol.12, No.24, (December 2002), pp. 2076-2089.

[52] Hemann, M. T. & Greider, C. W. (1999). G-Strand Overhangs on Telomeres in Telomerase-Deficient Mouse Cells. *Nucleic Acids Research*, Vol.27, No.20, (October 1999), pp. 3964-3969.

[53] Henderson, E. R. & Blackburn, E. H. (1989). An Overhanging 3' Terminus is a Conserved Feature of Telomeres. *Molecular and Cellular Biology*, Vol.9, No.1, (January 1989), pp. 345-348.

[54] Henson, J. D.; Neumann, A. A., Yeager, T. R. & Reddel, R. R. (2002). Alternative Lengthening of Telomeres in Mammalian Cells. *Oncogene*, Vol.21, No.4, (January 2002), pp. 598-610.

[55] Hsu, H. L.; Gilley, D., Blackburn, E. H. & Chen, D. J. (1999). Ku is Associated with the Telomere in Mammals. *Proceedings of the National Academy of Sciences of the United States of America*, Vol.96, No.22, (October 1999), pp. 12454-12458.

[56] Hsu, H. L.; Gilley, D., Galande, S. A., Hande, M. P., Allen, B., Kim, S. H., Li, G. C., Campisi, J., Kohwi-Shigematsu, T. & Chen, D. J. (2000). Ku Acts in a Unique Way at the Mammalian Telomere to Prevent End Joining. *Genes and Development*, Vol.14, No. 22, (November 2000), pp. 2807-2812.

[57] Huffman, K. E.; Levene, S. D., Tesmer, V. M., Shay, J. W. & Wright, W. E. (2000). Telomere Shortening is Proportional to the Size of the G-Rich Telomeric 3'-Overhang. *The Journal of Biological Chemistry*, Vol.275, No.26, (June 2000), pp. 19719-19722.

[58] Kawanishi, S. & Oikawa, S. (2004). Mechanism of Telomere Shortening by Oxidative Stress. *Annals of the New York Academy of Sciences*, Vol.1019, (June 2004), pp. 278-284.

[59] Keys, B.; Serra, V., Saretzki, G. & Von Zglinicki, T. (2004). Telomere Shortening In Human Fibroblasts is not Dependent on the Size of the Telomeric-3'-Overhang. *Aging Cell*, Vol.3, No.3, (June 2004), pp. 103-109.

[60] Kim, S. H.; Kaminker, P. & Campisi, J. (1999). TIN2, a New Regulator of Telomere Length in Human Cells. *Nature Genetics*, Vol.23, No.4, (December 1999), pp. 405-412.

[61] Kirk, K. E.; Harmon, B. P., Reichardt, I. K., Sedat, J. W. & Blackburn E. H. (1997). Block in Anaphase Chromosome Separation Caused by a Telomerase Template Mutation. *Science*, Vol.275, No. 5305, (March 1997), pp. 1478-1481.

[62] Klobutcher, L. A.; Swanton, M. T., Donini, P. & Prescott, D. M. (1981). All Gene-Sized DNA Molecules in Four Species of Hypotrichs have the Same Terminal Sequence

and an unusual 3' Terminus. *Proceedings of the National Academy of Sciences of the United States of America*, Vol.78, No.5, (May 1981), pp. 3015-3019.

[63]   Krokan, H. E.; Standal, R., & Slupphaug, G. (1997). DNA Glycosylases in the Base Excision Repair of DNA. *Biochemical Journal*, Vol.325, No.Pt 1, (July 1997), pp. 1-16.

[64]   Kruk, P. A.; Rampino, N. J. & Bohr, V. A. (1995). DNA Damage and Repair in Telomeres: Relation to Aging. *Proceedings of the National Academy of Sciences of the United States of America*, Vol.92, No.1, (January 1995), pp. 258-262.

[65]   Kurenova, E. V. & Mason, J. M. (1997). Telomere Functions. A Review. *Biochemistry (Moscow)*, Vol.62, No.11, (November 1997), pp. 1242-1253.

[66]   Larrivee, M.; LeBel, C. & Wellinger, R. J. (2004). The Generation of Proper Constitutive G-Tails on Yeast Telomeres is Dependent on the MRX Complex. *Genes and Development*, Vol.18, No.12, (June 2004), pp. 1391-1396.

[67]   Lee, J. H. & Paull, T. T. (2007). Activation and Regulation of ATM Kinase Activity in Response to DNA Double-Strand Breaks. *Oncogene*, Vol.26, No.56, (December 2007), pp. 7741-7748.

[68]   Lenain, C.; Bauwens, S., Amiard, S., Brunori, M., Giraud-Panis, M. J. & Gilson, E. (2006). The Apollo 5' Exonuclease Functions Together with TRF2 to Protect Telomeres from DNA Repair. *Current Biology*, Vol.16, No.13, (July 2006), pp. 1303-1310.

[69]   Levy, M. Z.; Allsopp, R. C., Futcher, A. B., Greider, C. W. & Harley, C. B. (1992). Telomere End-Replication Problem and Cell Aging. *Journal of Molecular Biology*, Vol.225, No.4, (June 1992), pp. 951-960.

[70]   Li, B.; Oestreich, S. & de Lange, T. (2000). Identification of Human Rap1: Implications for Telomere Evolution. *Cell*, Vol.101, No.5, (May 2000), pp. 471-483.

[71]   Liang, L.; Deng, L., Nguyen, S. C., Zhao, X., Maulion, C. D., Shao, C. & Tischfield, J. A. (2008). Human DNA Ligases I and III, but not Ligase IV, are required for Microhomology-Mediated End Joining of DNA Double-Strand Breaks. *Nucleic Acids Research*, Vol.36, No.10, (June 2008), pp. 3297-3310.

[72]   Lieber, M. R.; Ma, Y., Pannicke, U. & Schwarz, K. (2003). Mechanism and Regulation of Human Non-Homologous DNA End-Joining. *Nature Reviews Molecular Cell Biology*, Vol.4, No.9, (September 2003), 712-720.

[73]   Lingner, J.; Cooper, J. P. & Cech, T. R. (1995). Telomerase and DNA End Replication: No Longer a Lagging Strand Problem? *Science*, Vol.269, No.5230, (September 1995), pp. 1533-1534.

[74]   Liu, L.; Trimarchi, J. R., Smith, P. J. & Keefe, D. L. (2002). Mitochondrial Dysfunction Leads to Telomere Attrition and Genomic Instability. *Aging Cell*, Vol.1, No.1, (October 2002), pp. 40-46.

[75] Londoño-Vallejo, J. A. (2008). Telomere Instability and Cancer. *Biochimie*, Vol.90, No. 1, (January 2008), pp. 73-82.

[76] Mackenney, V. J.; Barnes, D. E. & Lindahl, T. (1997). Specific Function of DNA Ligase I in Simian Virus 40 DNA Replication by Human Cell-Free Extracts is Mediated by the Amino-Terminal Non-Catalytic Domain. *The Journal of Biological Chemistry*, Vol. 272, No.17, (April 1997), pp. 11550-11556.

[77] Makarov, V. L.; Hirose, Y. & Langmore, J. P. (1997). Long G Tails at Both Ends of Human Chromosomes Suggest a C Strand Degradation Mechanism for Telomere Shortening. *Cell*, Vol.88, No.5, (March 1997), pp. 657-666.

[78] Maringele, L. & Lydall, D. (2002). EXO1-Dependent Single-Stranded DNA at Telomeres Activates Subsets of DNA Damage and Spindle Checkpoint Pathways in Budding Yeast yku70Delta Mutants. *Genes and Development*, Vol.16, No.15, (August 2002), pp. 1919-1933.

[79] McElligott, R. & Wellinger, R. J. (1997). The Terminal DNA Structure of Mammalian Chromosomes. *The EMBO Journal*, Vol.16, No.12, (June 1997), pp. 3705-3714.

[80] Meeker, A. K. & Coffey, D. S. (1997). Telomerase: A Promising Marker of Biological Immortality of Germ, Stem, and Cancer Cells. A Review. *Biochemistry (Moscow)*, Vol. 62, No.11, (November 1997), pp. 1323-1331.

[81] Mimitou, E. P. & Symington, L. S. (2009). DNA End Resection: Many Nucleases Make Light Work. *DNA Repair (Amsterdam)*, Vol.8, No.9, (September 2009), pp. 983-995.

[82] Muntoni, A. & Reddel R. R. (2005). The First Molecular Details of ALT in Human Tumor Cells. *Human Molecular Genetics*, Vol. 14, No.2, (October 2005), pp. 191-196.

[83] Murr, R.; Loizou, J. I., Yang, Y. G., Cuenin, C., Li, H., Wang, Z. Q. & Herceg, Z. (2006). Histone Acetylation by Trrap-Tip60 Modulates Loading of Repair Proteins and Repair of DNA Double-Strand Breaks. *Nature Cell Biology*, Vol.8, No.1, (January 2006), pp. 91-99.

[84] Nam, E. A. & Cortez, D. (2011). ATR Signalling: More than Meeting at the Fork. *Biochemical Journal*, Vol.436, No.3, (June 2011), pp. 527-536.

[85] Nosek, J.; Dinouël, N., Kovac, L. & Fukuhara, H. (1995). Linear Mitochondrial DNAs from Yeasts: Telomeres with Large Tandem Repetitions. *Molecular and General Genetics*, Vol.247, No.1, (April 1995), pp. 61-72.

[86] Nugent, C. I., Bosco, G., Ross, L. O., Evans, S. K., Salinger, A. P., Moore, J. K., Haber, J. E. & Lundblad, V. (1998). Telomere Maintenance is Dependent on Activities Required for End Repair of Double-Strand Breaks. *Current Biology*, Vol.8, No.11, (May 1998), p. 657-660.

[87] O'Connor, M. S.; Safari, A., Liu, D., Qin, J. & Songyang, Z. (2004). The Human Rap1 Protein Complex and Modulation of Telomere Length. *The Journal of Biological Chemistry*, Vol.279, No.27, (July 2004), pp. 28585-28591.

[88] O'Connor, M. S.; Safari, A., Xin, H., Liu, D. & Songyang, Z. (2006). A Critical Role for TPP1 and TIN2 Interaction in High-Order Telomeric Complex Assembly. *Proceedings of the National Academy of Sciences of the United States of America*, Vol.103, No.32, (August 2006), pp. 11874-11879.

[89] Ohki, R.; Tsurimoto, T. & Ishikawa, F. (2001). In Vitro Reconstitution of the End Replication Problem. *Molecular and Cellular Biology*, Vol.21, No.17, (September 2001), pp. 5753-5766.

[90] Olovnikov, A. M. (1971). Principle of Marginotomy in Template Synthesis of Polynucleotides. *Doklady Akademiii Nauk SSSR*, Vol.201, No.6, (n.d.), pp. 1496-1499.

[91] Olovnikov, A. M. (1973). A Theory of Marginotomy: The Incomplete Copying of Template Margin in Enzymatic Synthesis of Polynucleotides and Biological Significance of the Phenomenon. *Journal of Theoretical Biology*, Vol.41, No.1, (September 1973), pp. 181-190.

[92] Olovnikov, A. M. (1995a). The Effect of the Incomplete Terminal Repair of the Linear Double-Stranded DNA Molecule. *Izvestiia Akademii Nauk Seriia Biologicheskaia*, No.4, (July-August 1995), pp. 501-503.

[93] Olovnikov, A. M. (1995b). The Role of Incomplete Terminal Repair of Chromosomal DNA in the Aging of Neurons and Postmitotic Organisms. *Izvestiia Akademii Nauk Seriia Biologicheskaia*, No.4, (July-August 1995), pp. 504-507.

[94] Olovnikov, A. M. (1995c). The Possible Cellular Use of the Effect of Incomplete DNA Terminal Repair to Control the Correct Sequence of Events in Individual Development. *Ontogenez*, Vol.26, No.3, (May-June 1995), pp. 254-256.

[95] Palm, W. & de Lange, T. (2008). How Shelterin Protects Mammalian Telomeres. *Annual Reviews of Genetics*, Vol.42, (December 2008), pp. 301-334.

[96] Passos, J. F. & Von Zglinicki, T. (2006). Oxygen Free Radicals in Cell Senescence: Are they Signal Transducers? *Free Radical Research*, Vol.40, No.12, (December 2006), pp. 1277-1283.

[97] Passos, J. F.; von Zglinicki, T. & Saretzki, G. (2006). Mitochondrial Dysfunction and Cell Senescence: Cause Or Consequence? *Rejuvenation Research*, Vol.9, No.1, (Spring 2006), pp. 64-68.

[98] Passos, J. F.; Saretzki, G. & von Zglinicki, T. (2007). DNA Damage in Telomeres and Mitochondria during Cellular Senescence: is there a Connection? *Nucleic Acids Research*, Vol.35, No.22, (December 2007), pp. 7505-7513.

[99] Pedram, M.; Sprung, C. N., Gao, Q., Lo, A. W. I., Reynolds, G. E. & Murnane, J. P. (2006). Telomere Position Effect and Silencing of Transgenes near Telomeres in the Mouse. *Molecular and Cellular Biology*, Vol.26, No.5, (March 2006), pp. 1865-1878.

[100] Pennaneach, V.; Putnam, C. D. & Kolodner, R. D. (2006). Chromosome Healing by *de novo* Telomere Addition in *Saccharomyces cerevisiae*. *Molecular Microbiology*, Vol.59, No.5, (March 2006), pp. 1357-1368.

[101] Petersen, S.; Saretzki, G. & von Zglinicki, T. (1998). Preferential Accumulation of Single-Stranded Regions in Telomeres of Human Fibroblasts. *Experimental Cell Research*, Vol.239, No.1, (February 1998), pp. 152-160.

[102] Podgornaya, O. I.; Bugaeva, E. A., Voronin, A. P., Gilson, E. & Mitchell, A. R. (2000). Nuclear Envelope Associated Protein That Binds Telomeric DNAs. *Molecular Reproduction and Development*, Vol.57, No.1, (September 2000), pp. 16-25.

[103] Polotnianka, R. M.; Li, J. & Lustig, A. J. (1998). The Yeast Ku Heterodimer is Essential for Protection of the Telomere Against Nucleolytic and Recombinational Activities. *Current Biology*, Vol.8, No.14, (July 1998), pp. 831-834.

[104] Polyak, K.; Xia, Y., Zweier, J. L., Kinzler, K. W. & Vogelstein, B. (1997). A Model for p53-Induced Apoptosis. *Nature*, Vol.389, No.6648, (September 1997), pp. 300-305.

[105] Reardon, J. T. & Sancar, A. (2005). Nucleotide Excision Repair. *Progress in Nucleic Acid Research and Molecular Biology*, Vol.79, (n.d.), pp. 183-235.

[106] Rhyu, M. S. (1995). Telomeres, Telomerase, and Immortality. *Journal of the National Cancer Institute*, Vol.87, No.12, (June 1995), pp. 884-894.

[107] Richter, T.; Saretzki, G., Nelson, G., Melcher, M., Olijslagers, S. & von Zglinicki, T. (2007). TRF2 Overexpression Diminishes Repair of Telomeric Single-Strand Breaks and Accelerates Telomere Shortening in Human Fibroblasts. *Mechanisms of Ageing and Development*, Vol.128, No.4, (April 2007), pp. 340-345.

[108] Richter, T. & von Zglinicki, T. (2007). A Continuous Correlation between Oxidative Stress and Telomere Shortening in Fibroblasts. *Experimental Gerontology*, Vol.42, No. 11, (November 2007), pp. 1039-1042.

[109] Sancar, A.; Lindsey-Boltz, L. A., Unsal-Kaçmaz, K. & Linn, S. (2004). Molecular Mechanisms of Mammalian DNA Repair and the DNA Damage Checkpoints. *Annual Review of Biochemistry*, Vol.73, (July 2004), pp. 39-85.

[110] Saretzki, G.; Sitte, N., Merkel, U., Wurm, R. E. & von Zglinicki, T. (1999). Telomere Shortening Triggers a p53-dependent Cell Cycle Arrest via Accumulation of G-rich Single Stranded DNA Fragments. *Oncogene*, Vol.18, No.37, (September 1999), pp. 5148-5158.

[111] Seeberg, E.; Eide, L. & Bjørås, M. (1995). The Base Excision Repair Pathway. *Trends in Biochemical Sciences*, Vol.20, No.10, (October 1995), pp. 391-397.

[112] Sfeir, A. J.; Chai, W., Shay, J. W. & Wright, W. E. (2005). Telomere-End Processing: the Terminal Nucleotides of Human Chromosomes. *Molecular Cell*, Vol.18, No.1, (April 2005), pp. 131-138.

[113] Sohn, S. Y. & Cho, Y. (2009). Crystal Structure of the Human Rad9-Hus1-Rad1 Clamp. *Journal of Molecular Biology*, Vol.390, No.3, (July 2009), pp. 490-502.

[114] Stansel, R. M.; de Lange, T. & Griffith, J. D. (2001). T-Loop Assembly *in vitro* Involves Binding of TRF2 near the 3' Telomeric Overhang. *The EMBO Journal*, Vol.20, No.19, (October 2001), pp. 5532-5540.

[115] Stewart, S. A.; Ben-Porath, I., Carey, V. J., O'Connor, B. F., Hahn, W. C. & Weinberg, R. A. (2003). Erosion of the Telomeric Single-Strand Overhang at Replicative Senescence. *Nature Genetics*, Vol. 33, No.4, (April 2003), pp. 492-496.

[116] Stewart, S. A. & Weinberg, R. A. (2006). Telomeres: Cancer to Human Aging. *Annual Rewiev of Cell and Developmental Biology*, Vol.22, (November 2006), pp. 531-557.

[117] Stewart, S. A. (2005). Telomere Maintenance and Tumorigenesis: An "ALT"ernative Road. *Current Molecular Medicine*, Vol.5, No.2, (March 2005), pp. 253-257.

[118] Tarsounas, M.; Muñoz, P., Claas, A., Smiraldo, P. G., Pittman, D. L., Blasco, M. A. & West, S. C. (2004). Telomere Maintenance Requires the RAD51D Recombination/ Repair Protein. *Cell*, Vol.117, No.3, (April 2004), pp. 337-347.

[119] Tchirkov, A. & Lansdorp, P. M. (2003). Role of Oxidative Stress in Telomere Shortening in Cultured Fibroblasts from Normal Individuals and Patients with Ataxia-Telangiectasia. *Human Molecular Genetics*, Vol.12, No.3, (February 2003), pp. 227-232.

[120] Testorelli, C. (2003). Telomerase And Cancer. *Journal of Experimental and Clinical Cancer Research*, Vol.22, No.2, (June 2003), pp. 165-169.

[121] Toussaint, O.; Medrano, E. E. & von Zglinicki, T. (2000). Cellular and Molecular Mechanisms of Stress-Induced Premature Senescence (SIPS) of Human Diploid Fibroblasts and Melanocytes. *Experimental Gerontology*, Vol.35, No.8, (October 2000), pp. 927-945.

[122] Tran, P. T.; Erdeniz, N., Symington, L. S. & Liskay, R. M. (2004). EXO1 – A Multi-Tasking Eukaryotic Nuclease. *DNA Repair (Amsterdam)*, Vol.3, No.12, (December 2004), pp. 1549-1559.

[123] van Overbeek, M. & de Lange, T. (2006). Apollo, an Artemis-Related Nuclease, Interacts with TRF2 and Protects Human Telomeres in S Phase. *Current Biology*, Vol.16, No.13, (July 2006), pp. 1295-1302.

[124] von Zglinicki, T.; Saretzki, G., Döcke, W. & Lotze, C. (1995). Mild Hyperoxia Shortens Telomeres and Inhibits Proliferation of Fibroblasts: a Model for Senescence? *Experimental Cell Research*, Vol.220, No.1, (September 1995), pp. 186-193.

[125] von Zglinicki, T. (2000). Role of Oxidative Stress in Telomere Length Regulation and Replicative Senescence. *Annals of the New York Academy of Sciences*, Vol.908, (June 2000), pp. 99-110.

[126] von Zglinicki, T.; Pilger, R. & Sitte, N. (2000). Accumulation of Single-Strand Breaks is the Major Cause of Telomere Shortening in Human Fibroblasts. *Free Radical Biology and Medicine*, Vol.28, No.1, (January 2000), pp. 64-74.

[127] von Zglinicki, T. (2002). Oxidative Stress Shortens Telomeres. *Trends in Biochemical Sciences*, Vol.27, No.7, (July 2002), pp. 339-344.

[128] Watson, J. D. (1972). Origin of Concatemeric T7 DNA. *Nature New Biology*, Vol.239, No.94, (October 1972), pp. 197-201.

[129] Wellinger, R. J.; Wolf, A. J. & Zakian, V. A. (1993). Saccharomyces Telomeres Acquire Single-Strand TG1-3 Tails Late in S Phase. *Cell*, Vol.72, No.1, (January 1993), pp. 51-60.

[130] Williams, G. J.; Lees-Miller, S. P. & Tainer, J. A. (2010). Mre11-Rad50-Nbs1 Conformations and the Control of Sensing, Signaling, and Effector Responses at DNA Double-Strand Breaks. *DNA Repair (Amsterdam)*, Vol.9, No.12, (December 2010), pp. 1299-1306.

[131] Wright, W. E.; Tesmer, V. M., Huffman, K. E., Levene, S. D. & Shay, J. W. (1997). Normal Human Chromosomes Have Long G-rich Telomeric Overhangs at One End. *Genes and Development*, Vol.11, No.21, (November 1997), pp. 2801-2809.

[132] Wu, P.; van Overbeek, M., Rooney, S. & de Lange, T. (2010). Apollo Contributes to G Overhang Maintenance and Protects Leading-End Telomeres. *Molecular Cell*, Vol.39, No.4, (August 2010), pp. 606-617.

[133] Ye, J. Z.; Donigian, J. R., van Overbeek, M., Loayza, D., Luo, Y., Krutchinsky, A. N., Chait, B. T. & de Lange T. (2004). TIN2 Binds TRF1 and TRF2 Simultaneously and Stabilizes the TRF2 Complex on Telomeres. *The Journal of Biological Chemistry*, Vol. 279, No.45, (November 2004), pp. 47264-47271.

[134] Yuan, X.; Ishibashi, S., Hatakeyama, S., Saito, M., Nakayama, J., Nikaido, R., Haruyama, T., Watanabe, Y., Iwata, H., Iida, M., Sugimura, H., Yamada, N. & Ishikawa, F. (1999). Presence of Telomeric G-Strand Tails in the Telomerase Catalytic Subunit TERT Knockout Mice. *Genes to Cells*, Vol.4, No.10, (October 1999), pp. 563-572.

[135] Zheng, L. & Shen, B. (2011). Okazaki Fragment Maturation: Nucleases Take Centre Stage. *Journal of Molecular Cell Biology*, Vol.3, No.1, (February 2011), pp. 23-30.

[136] Zhu, X. D.; Küster, B., Mann, M., Petrini, J. H. & de Lange, T. (2000). Cell-Cycle-Regulated Association of RAD50/MRE11/NBS1 with TRF2 and Human Telomeres. *Nature Genetics*, Vol.25, No.3, (July 2000), pp. 347-352.

[137]   Zou, L. & Elledge, S. J. (2003). Sensing DNA Damage through ATRIP Recognition of RPA-ssDNA Complexes. *Science*, Vol.300, No.5625, (June 2003), pp. 1542-1548.

[138]   Zubko, M. K.; Guillard, S. & Lydall, D. (2004). Exo1 and Rad24 Differentially Regulate Generation of ssDNA at Telomeres of *Saccharomyces cerevisiae* cdc13-1 Mutants. *Genetics*, Vol.168, No.1, (September 2004), pp. 103-115

Chapter 9

Telomeres: Their Structure and Maintenance

Radmila Capkova Frydrychova and James M. Mason

Additional information is available at the end of the chapter

1. Introduction

Telomeres are essential nucleoprotein structures at the ends of eukaryotic chromosomes. They play several essential roles preserving genome stability and function, including distinguishing chromosome ends from DNA double stranded breaks (DSBs) and maintenance of chromosome length. Due to the inability of conventional DNA polymerases to replicate the very end of a chromosome, sometimes known as the end replication problem, chromosome ends shorten with every round of DNA replication. In the absence of special telomere maintenance mechanisms this telomere shortening leads to replicative senescence and apoptosis. Several telomere maintenance mechanisms have been identified; these are reflected in several known types of telomeres. In most eukaryotes telomeres comprise a tandem array of a short, 5-8 bp, well conserved repeat unit, and telomere length is maintained by telomerase, a specialized reverse transcriptase that carries its own RNA template and adds telomeric sequences onto chromosome ends [1]. Nevertheless, in some organisms the array of short telomeric sequence motifs has been replaced with less conventional sequences, such as satellite sequences or transposable elements. The telomeres of such organisms are maintained through homologous recombination or through transposition of the mobile elements [2,3]. These different telomere types present distinct difficulties for chromosome end protection. Telomeres maintained by telomerase are protected by a proteinaceous telomere cap, termed shelterin, that recognizes chromosome ends in a DNA sequence specific manner, while telomeres with long terminal repeat units are protected by a cap, termed terminin, that binds to chromosome ends independently of DNA sequence.

2. The structure of telomeric DNA: "usual" and "unusual" telomeres

The most common telomere structure found across the whole eukaryotic tree is a simple telomeric repeat of the form $(T_xA_yG_z)_n$ generated by telomerase. For example, the sequence in unikonts generally, including animals, fungi and amoebozoa, is T_2AG_3, while in most plants and green algae it is T_3AG_3. Within these broad generalizations, however, there are exceptions. Some species seem to have lost the canonical telomeric motif altogether. We will mention a few examples here, then describe one of these examples in more detail.

2.1. Chromalveolata

The terminal sequence motif seems to be quite variable among the Chromalveolates, while still adhering to the consensus telomeric motif (Figure 1). Apicomplexa species use three different motifs [4-6], and ciliates use two [4,7]. Dinoflagellates use T_3AG_3 [8], similar to plants and green algae, while diatoms use T_2AG_3 [9], similar to unikonts. Photosynthetic species in the Chromalveolates are derived from the engulfment of a red alga. The resulting nucleomorphs retain the algal linear chromosomes and telomeres that are very different. The cryptomonad, *Guillardia theta*, for example, uses T_3AG_3 in its nucleus and $(AG)_7A_2G_6A$ in its nucleomorph [11,12].

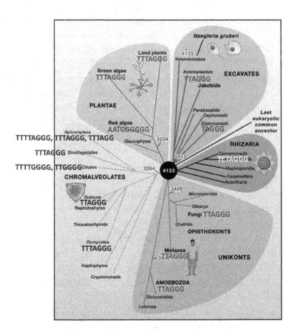

Figure 1. Diagram showing five major eukaryotic supergroups and representative telomeric motifs. These groups are shown to have diverged from a single latest common ancestor, because the evolutionary relationships are not known. Trees connecting the major taxa within these supergroups are shown, but the branch lengths are arbitrary. Representative telomeric motifs are shown for the major subtaxa. In some cases, two or three representative motifs are known for one of these taxa, as shown. Exceptions to these representations are discussed in the text. The figure was modified from [10] with permission.

It seems likely that the telomere binding proteins in these organisms are either different in the two intracellular bodies, or do not bind in a DNA sequence specific manner.

2.2. Plantae

Among the Plantae, land plants and green algae mostly use T_3AG_3 as a telomeric motif, while the red algae have a very different sequence at their chromosome ends. The red alga *Cyanidioschyzon merolae*, for example, uses A_2TG_6 [13]. While telomeres in most green algae conform to the telomeric motif of this kingdom, the order Chlamydomonadales includes species that carry the telomeric motifs T_4AG_3, T_3AG_3 and T_2AG_3, apparently independently of phylogeny as determined by the 18S rDNA sequence [14]. Some species of this order lack all three of these telomeric motifs and carry unknown DNA sequences at their chromosome ends. It is possible that the 18S rDNA sequence does not represent an accurate reflection of phylogeny or the telomeric motif is quite variable in this order. In either case, it seems that sequence specific binding by telomeric proteins may have eased in this order.

Similarly, while most land plants retain the canonical T_3AG_3 telomeric motif, telomeres in a few orders differ from this structure. Within the monocot order Asparagales some species of Alliaceae have switched to the sequence T_2AG_3, and others appear to have lost the canonical telomeric sequence completely. It has been proposed that the telomeres of these latter Alliaceae species are maintained through transposition of mobile elements or through homologous recombination between the satellite sequences [15,16]. In the eudicot order Solanales the canonical telomeric motif as well as telomerase are absent from several genera of the family Solanaceae [17-20]. The actual telomeric sequence and compensation mechanism in this group of plants, however, remain unknown.

2.3. Unikonta

The T_2AG_3 telomeric sequence is found widely among the unikonts (Figure 1). While this is generally true within the fungi, representatives of two classes, Schizosaccharomycetes and Saccharomycetes, use variable, degenerate telomeric sequences that may result from replication infidelity or slippage [12]. In *Saccharomyces cerevisiae*, for example, the repeat motif is TG_{1-3}.

Similarly, T_2AG_3 is found widely at chromosome ends among metazoans. The animal phylum Arthropoda, however, uses the sequence T_2AG_2 at telomeres, and its sister phylum Tardigrada lacks both of these telomeric motifs [21]. Insects are the largest class of arthropods, and even here individual insect taxa may have different forms of the canonical sequence or even unrelated telomeric sequences. Insects seem to have lost the canonical arthropod telomeric motif several times [22,23]. In some cases, such as the coleopteran superfamily Tenebrionoidea, the arthropod repeat has been replaced by a similar motif, in this case $TCAG_2$ [24], while in many other instances the new telomeric DNA sequence has not been identified.

Insects of the orders Diptera, Mecoptera and Siphonaptera (superorder Antliophora) do not carry a canonical telomeric DNA sequence at their chromosome ends [23,25]; nor do they have a telomerase gene [26], indicating that telomerase was lost some 260-280 Mya. Even so, Diptera is one of the most successful insect orders, with some 152,000 species [27]. This suggests that telomerase and the canonical telomeric DNA sequences generated by telomerase, *per se*, are not critical for evolutionary survival. It is possible telomerase is expendable, as long as the telomere capping complex is compatible with whatever terminal DNA sequence is present on chromosome ends. When the primary pathway for telomere replication is defective, an alternative backup mechanism can restore telomere function. It was documented in yeast. Yeast mutants lacking telomerase showed the progressive telomere loss and, although the majority of the cells died, a minor subpopulation survived via homologous recombination [28].

Long satellite sequences have been reported in nematoceran species. Chromosome tips of several *Chironomus* species (infraorder Culicomorpha) consist of large, 50-200 kb, blocks of complex, tandemly repeated sequences that have been classified into subfamilies based on sequence similarities. Different telomeres display different sets of subfamilies, and the distribution of subfamilies differs between individuals within a species. The variation of the satellite sequences supports the proposal that telomeres in *Chironomus* are elongated by a gene conversion mechanism involving these long blocks of complex repeat units [29-32]. A similar situation has been found in *Anopheles gambiae* (infraorder Culicomorpha) using a plasmid fortuitously inserted into the complex telomeric sequences at the tip of chromosome 2L. The telomere carrying the plasmid was found to engage in frequent recombination events that resulted in extension of the terminal array [33,34]. Recently, a similar case was reported in *Rhynchosciara americana* (infraorder Bibionomorpha). Tandem arrays of short repeats, 16 and 22 bp in length, were found to extend to chromosome ends [35]. Although telomere elongation could not be assayed in this case, it seems likely that the mechanism is similar to that seen in other nematoceran species. In many respects, these complex arrays resemble subtelomeric sequences [36], suggesting a possible mechanism for telomere formation upon the loss of telomerase.

Telomere structures have only been examined in a single brachyceran genus, *Drosophila* (infraorder Muscomorpha). Telomeric DNA sequences consist of long arrays of non-long terminal repeat (LTR) retrotransposons and are thus very different from those found in Nematocera. These telomeric transposons resemble long interspersed elements (LINEs) found in mammals, but have some differences that may reflect their telomere-specific 'lifestyle.' Three families of telomeric elements have been described in *Drosophila melanogaster* (subgenus Sophophora), *HeT-A, TART* and *TAHRE* [2]; in all cases these elements are attached to the chromosome by their 3' oligo(A) tails. Many of the elements are truncated at the 5' end, possibly due to the end replication problem. *HeT-A* transposons are about 6 kb in length and make up about 80-90% of the elements found at chromosome ends. They are atypical LINE-like elements in three respects: the 3' untranslated region (UTR) comprises about 3 kb or half the length of the element; the transcriptional promoter is at the 3' end of the element to prevent loss when the element is present at the chromosome terminus with its 5' end exposed to incomplete DNA replication of linear DNA; and an open reading frame (ORF) coding for a reverse transcriptase is absent. *TART* elements are about 10 kb in length and make up about 10-20% of the telomeric retrotransposons. They are also unusual elements, but in some ways that differ from *HeT-A*: they also have an unusually long 3' UTR; they have a relatively strong antisense promoter of unknown function and a pair of perfect long non-terminal repeats that may be important for replication [37,38]; they make a reverse transcriptase, but the encoded Gag-like protein is unable to target telomeres in the absence of the *HeT-A* Gag [39]. *TAHRE* elements closely resemble *HeT-A*, except they encode their own reverse transcriptase. Thus, while *TAHRE* seems to be the only one of the three elements capable of independent transposition, it is by far the least abundant, comprising only 1% of the telomeric retrotransposons.

HeT-A and *TART* elements have also been found in *Drosophila virilis* (subgenus Drosophila) Although there is little sequence homology across species, the two types of retrotransposons can be recognized by their telomeric locations and unusual structures, as described above [40,41]. Given the difficulty in finding homology between evolutionarily related telomeric elements within the *Drosophila* genus, finding similar elements in other brachyceran species based on homology alone is unlikely. Thus, it is not known when these targeted transpositions took over the role of telomere maintenance from homologous recombination.

Human telomeres have been shown to form a large terminal loop dependent on the presence of a 3' G strand overhang at the telomeric end. This 3' end is tucked back into the double-stranded DNA as a loop, termed a t-loop [42]. Similar t-loops may also be formed in yeast [43].

3. Proteins associated with telomeres

The telomere cap, a multiprotein structure at chromosome end ensuring stability and integrity of the genome, was revealed by early cytological observations of chromosomal rearrangements after exposure to ionizing radiation [44]. The telomere cap allows cells to distinguish their natural chromosome ends from DSBs, thus protecting the chromosome termini from inadvertent DNA damage response (DDR) activities. Defects in the cap, or DSBs elsewhere in the genome, lead to activation of cell cycle checkpoints followed by DDR mechanisms. A consequence of inappropriate DSB repair are end-to-end fusions of chromosomes, i.e. formation of ring chromosomes or dicentric linear chromosomes, followed by chromosome breakage, which results in genomic instability and loss of cellular viability [45,46]. Although, in this context, telomeres perform the same essential function across phyla, cap proteins of diverse organisms are less conserved that one might expect. Even within a single taxonomic class, such as mammals, telomeric proteins display less conservation than other chromosomal proteins [47]. In mammals the telomere-specific cap complex has been termed 'shelterin' (Figure 2). The six-protein complex is formed by double-stranded TTAGG repeat-binding proteins TRF1 and TRF2, which recruit TIN2 and TPP1. The latter proteins make a bridge between the TRF proteins and G-overhang-binding protein, POT1. The sixth protein is the TRF2-interacting protein RAP1 [46,48,49]. A characteristic of shelterin proteins is specific and exclusive association with telomeric DNA, where they are permanently present throughout the cell cycle and serve as platform for a transient and dynamic recruitment of a number of telomere-associated factors, referred to as non-shelterin telomeric proteins. These non-shelterin proteins are required for telomere protection and replication but also have numerous nontelomeric functions. Examples include DDR proteins that are commonly involved in DSB repair through nonhomologous end joining (NHEJ) or homologous recombination (HR), such as ATM, ATR and Ku70/80, which associate with TRF1 and TRF2, and the MRN complex, composed of the MRE11, RAD50 and NBS1 (MRN) proteins, which associates with TRF2 [50-55]. Another protein associated with TRF2 is Apollo, an exonuclease important for recreating the 3' overhang [51,56]. The binding of shelterin proteins and formation of a functional cap require a terminal DNA array of specific sequence and of satisfactory length.

Analysis of deleterious events at shelterin-free telomeres revealed six pathways for end protection [57]. The primery protection by shelterin is against classical NHEJ and unwanted activation of ATM and ATR signaling. Additionally, shelterin provides a defense against alternate NHEJ, HR and 5' end resection. Another protective layer is achieved through the Ku70/80 heterodimer or 53BP1. 53BP1 minimizes resection but only at telomeres eliciting a DNA damage signal. Ku70/80 blocks alternate NHEJ and HR at telomeres independent of a DNA damage signal [57].

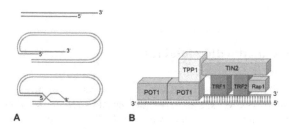

Figure 2. A. The telomere forms a t-loop structure characterized by invasion of the 3' overhang into a double stranded telomeric DNA. B. Six proteins, TRF1, TRF2, TPP1, POT1, TIN2, and RAP1 form a dedicated telomere-protection protein complex in humans [48,49,58].

Telomeres in *Saccharomyces cerevisiae* are protected by two separate protein complexes. One is the Rap1/Rif1/Rif2 complex, which localizes to double-stranded telomeric DNA. The other is the Cdc13/Stn1/Ten1 (CST) complex, which is targeted to the single-stranded G-overhangs through sequence-specific binding of Cdc13. Defects in the CST complex result in degradation of the C-stand and activation of DDR mechanisms [47]. As with shelterin, CST interacts with numerous proteins required for telomere function. Some evolutionary conservation in the protein composition of the cap is expected, for instance similarities to CST and shelterin are observed in telomeric proteins of numerous organisms. This is documented for mammalian CST, which, although not involved in telomere capping, facilitates telomere replication and, if impaired, leads to catastrophic telomeric defects [59]. Another example is Ver, a component of the *Drosophila* cap with weak structural similarities to Stn1 [60,61].

A multiprotein capping complex in *Drosophila*, termed 'terminin,' is an analog of mammalian shelterin [61]. One major difference between these two complexes is that terminin does not bind to a specific telomeric DNA sequence. Rather limited information is available about the structure and function of four known terminin proteins, HOAP, Moi, Ver, and HipHop. As with shelterin, terminin proteins localize specifically to telomeres and appear to function only at telomeres. HOAP is encoded by the *cav* gene [62]; Moi is a HOAP-binding protein encoded by *moi* [63,64]; Ver is structurally homologous to STN1 and is encoded by *ver* [60]; and HipHop is a HP1-HOAP interacting protein [65]. Assembly of the terminin complex requires strict dependencies. For example, the binding of HOAP and HipHoP to telomeres is interdependent, loss of one protein reduces binding of the other [65]; HOAP is required for Ver and Moi localization [61]. The terminin complex seems to occupy a broad region covering a more than 10 kb from the chromosome termini [65]. As with shelterin proteins, defects in terminin proteins lead to frequent telomeric fusions.

As there is no specific telomeric DNA sequence in Drosophila, terminin binding to telomeric DNA is sequence-independent, which makes a substantial difference between mammalian and Drosophila telomeres. In contrast to mammals, the complete loss of a Drosophila telomere does not definitely mean inescapable damage to genome stability and cell death, because under the right circumstances the telomere cap can be formed *de novo* as on any broken chromosome end and perform there the same protective functions as the regular telomere. This demonstrates that the telomeric retrotransposons, although important for telomere elongation, are not required as an unique assembly platform for cap formation [2,66,67].

Similar to shelterin, terminin presents a docking site for binding of additional proteins, called non-terminin capping proteins. Although not exclusively located at telomeres and having some telomere-unrelated functions, these proteins are required for the capping function and, in many cases, facilitate terminin assembly. There are several known non-terminin proteins; most of them were identified because their mutants display frequent telomeric fusions [61]. The best characterized is HP1a that is encoded by *Su(var)205*. The presence of HP1 at telomeres is required for HOAP binding, which reveals the importance of HP1 for terminin assembly. As in mammals, other non-terminin proteins are DNA repair factors: the *Drosophila* homologs of the ATM kinase and proteins of the MRN complex. Defects in the MRN complex lead to reduction of HOAP and HP1 at telomeres and frequent telomeric fusions. Through its effects on the binding of HOAP and possibly other terminin components, the MRN complex seems to be essential for the terminin formation [61,68,69]. ATM prevents telomeric fusions, and defects in this protein partially affect HP1/HOAP localization [70-72]. Woc is a zinc-finger protein preventing telomeric fusions, but acting independently of HP1, HOAP, and RAD50 [73]. UbcD1 is an E2 ubiquitin conjugating enzyme. It has been suggested that UbcD1-mediated ubiquitination of telomeric proteins is an essential post-translational modification ensuring their proper function [61,74]. In contrast to non-terminin and non-shelterin proteins that are largely conserved, a comparison between shelterin and terminin reveals no obvious homology in protein composition. Loss of conservation between shelterin and terminin proteins may correspond the evolutionary stage when a Antliophoran ancestor lost telomerase-based telomere elongation and had to evolve a sequence-independent protection of chromosome ends and acquire a new elongation system.

A highly condensed chromatin structure is a common characteristic of telomeres from yeast to man. Usually telomeres are heterochromatic, and the heterochromatic properties are thought to play an important role in telomeric function [75-77]. Telomeric chromatin is the source of telomeric position effect (TPE), a silencing of transgenes inserted into telomeres or their vicinity [78]. Besides the cap region, *Drosophila* telomeres contain two distinct chromatin domains: a subtelomeric region of repetitive DNA, termed TAS (telomere associated sequence), exhibiting features that resemble heterochromatin, and a terminal array of retrotransposons with euchromatic characteristics [79]. The *Drosophila* TAS region is, in contrast to retrotransposon array, the source of TPE [79,80]. Although organized into a heterochromatic structure, the vertebrate TTAGGG sequence remains unmethylated due to the lack of a appropriate cytosine substrate. The subtelomeric region is, in contrast, heavily methylated by DNA methyltranferases DNMT1, DNMT3a and DNMT3b [81]. Both in vertebrates and *Drosophila*, telomeric and subtelomeric regions are enriched in histone H3 methylated at lysine 9 (H3K9me), mediated by a H3K9-specific histone methyltransferase and HP1.

4. Telomeric replication and its difficulties

Based on DNA and protein composition, telomeres are typical heterochromatin, so their replication should correspond with a common paradigm of late heterochromatin replication. Based on early microscopic studies, it is generally accepted that DNA replication at early stages of S phase is associated with expressed genes, whereas repressed tissue-specific genes or heterochromatic regions are replicated during the late stages of replication [82-84]. The late replication seems to be common, but definitely is not universal [85]. Replication of human telomeres takes place throughout S phase, and specific telomeres tend to replicate at defined stages, some replicating early and others late [86]. The pattern of replication timing seems to be conserved between homologous chromosomes and does not vary between cells of different individuals. Although no correlation was found with telomere length or telomerase activity, a strong association was observed with nuclear localization. Late-replicating telomeres show a preferential association with the nuclear periphery, while early-replicating telomeres are preferentially located near the nuclear center [86]. A different situation was found in budding yeast, *Saccharomyces cerevisiae*, where early telomere replication correlates with short telomeric length and telomerase activity [87,88]. In fission yeast, *Schizosaccharomyces pombe*, telomere replication corresponds to S/G2 phase [85,89].

Because of the repetitive nature of telomeric DNA, telomeres present a significant problem for their replication. Spontaneous replication fork regression in telomeric DNA *in vitro* was determined to be 41% higher than seen in non-repeated DNA [90]. The obstacles during replication may lead to formation of cruciform intermediates, resulting in unwanted recombination events, amplifications or deletions [90,91]. Most of the telomere is replicated by a standard replication fork, however, to achieve efficient telomere replication a number of additional steps are needed. The process requires cooperation between standard replication factors and telomeric proteins, DDR proteins and numerous additional factors [47]. Examples of additional proteins are RecQ-type helicases that are present at replication forks in addition to standard helicases and are shown to unwind structures similar to chickenfoot intermediates [90,92]. Cooperation of replication factors with shelterin proteins is also documented. TRF1 mutants showed a reduction in replication efficiency, suggesting that TRF1 promotes efficient replication of telomeric DNA by preventing fork stalling [93]. Similarly, Taz1, a TRF1 homolog in fission yeast, has been shown to prevent fork stalling [94]. Another example is mammalian CTC1; deletion of CTC1 results in increased loss of leading C-strand telomeres, dramatic telomere loss and accumulation of excessive single-stranded telomeric DNA [95].

In yeast, the replication of telomeres is initiated in subtelomeric regions, and the replication fork moves towards the chromosome termini [96]. In mammalian cells, the origin of telomeric replication and direction is ambiguous.

After the replication fork reaches the chromosome terminus, the lagging strand gains a 3' overhang due to the removal of the primer for the terminal Okazaki fragment. At the same time C-strand specific resection occurs by nucleases Exo1 and/or Dna2 to produce a G-overhang on the leading strand [89]. If active, telomerase elongates the G-overhangs by addition of new telomeric repeats. Telomerase action is followed by complementary C-strand synthesis by DNA pol α. The process is terminated by additional processing to remove the RNA primer and to leave a 40-400 nucleotide G-overhang. The timing of the events differs between species. In human cells, telomere replication occurs at the same time as telomerase-mediated extension, and fill-in synthesis of C- strand is delayed until S/G2. Budding yeast shows tight coupling between G-strand extension and C-strand synthesis [89,91].

5. The mechanisms of telomeric elongation and their regulation

Telomerase is a ribonucleoprotein reverse transcriptase that utilizes its protein subunit (TERT in mammals, Est2p in *S. cerevisiae*) to elongate the 3' end of telomeric DNA using an internal RNA subunit (TR) as a template [97-99]. Telomerase activity is related to cell proliferation status: it is high in actively cycling cultures and low in quiescent differentiated cells [100]. Telomerase is not detected in human mature sperm or unfertilized eggs, but after fertilization telomerase is rapidly activated. A dramatic increase is observed in blastocysts, but during later stages of gestation telomerase activity declines. In the 16-week fetus Wright [101] showed high levels of telomerase in liver and intestine; detectable activity in lung, skin, muscle, adrenal glands, and kidney; and very weak or no activity in brain, bone or placental extracts. Most somatic cells in adults show no telomerase activity, as enzyme activity is limited to specific types of proliferative cells, such as embryonic, stem and epithelial cells, the germline, or cells of the hematopoetic system [102,103]. Telomerase activity is highly regulated. Reactivation of telomerase is associated with tumor development, and conversely, insufficient telomerase activity is linked to stem cell diseases, such as dyskeratosis congenita and aplastic anemia [104-106].

Telomerase is regulated through genetic, epigenetic and environmental factors: TERT and TR transcription, posttranscriptional and posttranslational modifications of TERT, and telomerase recruitment and processivity [104]. TERT promoter activity has been studied extensively, and numerous transcription factors have been found to interact with TERT. TERT transcription is, for instance, activated by the oncogene c-Myc and suppressed by the tumor-suppressor WT1 (Wilm's tumor suppresor). Misregulation of TERT through the c-Myc or WT1 pathways is associated with telomerase reactivation in cancer cells [107,108]. Although transcription of TERT is the major determinant of telomerase activity, TERT transcript levels do not always correlate with enzyme activity. Posttranslational phosphorylation may regulate telomerase activity, as may telomerase degradation through ubiquitination, as the half-life of telomerase activity was approximately 24 hours [109]. In human cells the POT1-TPP1 complex was found to be a key regulator of telomerase processivity [110-113].

Little is known about the regulation of telomere length in *Drosophila*, where two modes of telomere elongation have been described: transposition of telomeric elements and gene conversion. The process of telomeric transposition is composed of several steps: 1. transcription of the telomeric elements, 2. export of retroelement transcripts from the nucleus to the cytoplasm, 3. translation, 4. nuclear re-import of the retroelement transcripts together the retroelement proteins, 5. recognition of chromosome ends, and 6. target-primed reverse transcription, which attaches the 3' oligo(A) tails of the elements to chromosome termini [2]. Transposition of these elements to chromosome termini does not depend on a specific DNA sequence at the target site and together with the loss of telomeric DNA results in tandem arrays of mixed complete and 5' truncated elements [2]. The regulation of telomere elongation may be on the level of retroelement transcription and/or accessibility of the chromosome ends for new retroelement attachments. A variety of proteins have been identified to play a role in *Drosophila* telomere capping, however, only a few proteins are known to function in telomere elongation. HP1 was found to have a dual role in telomere protection and telomere length control. Compared to wild-type, heterozygotous *Su(var)205* mutants displayed much longer telomeres associated with a dramatic increases in retroelement transcription and transposition [114-116]. The regulation of retroelement transcription by HP1 was observed along the terminal retrotransposon array, thus this HP1 function is not limited to the telomere cap [117]. No, or only minor, changes were observed in telomere length or retroelement transcription in mutants of genes involved in telomere capping, such as *cav*, *moi*, *ver* or *atm* [61,117], which may indicate that terminin does not control telomere length. Another gene regulating telomere length is *prod*. Although *prod* mutants showed elevated levels of *HeT-A* transcripts, no change in telomere length was observed, suggesting that elevated retroelement transcription is not sufficient for telomere length growth [118]. Similar data were observed for members of rasiRNA (repeat-associated small interfering RNA) pathway *aub* (*aubergine*) and *Spn-E*. Their mutants displayed higher *HeT-A* transcript levels [119], albeit without any significant increase in telomere length (our unpublished data). In parallel with telomerase activity, transcription of telomeric elements is observed only in proliferating cells, such as embryonic cells, cells of imaginal discs, testis and ovaries [120,121].

Telomere length is maintained through an interplay between telomere maintenance mechanisms and telomere shortening events. Based on human research it has been proposed that telomerase activity and telomere length are modulated by different endogenous and exogenous factors, such as emotional or physical stress, health conditions and aging [102]. However, the prime factor in telomeric shortening may well be oxidative stress. Due to a high content of guanines, telomeres are particularly vulnerable to oxidative damage, and the impact of oxidative stress on telomeric length has been proposed to be much larger than the end-replication problem [122]. Endogenous oxidative stress is associated with several cellular processes, such as the mitochondrial OXPHOS system and inflammation. Mitochondrial dysfunction-induced reactive oxygen species and hyperoxia *in vitro* lead to accelerated telomere shortening and reduced proliferative lifespan of cultured somatic cells [123]. Thus, short telomeric length in humans appears to be linked to the limited proliferative capacity of normal somatic cells, and it is likely that telomeric shortening is one of the key events related to cellular senescence and organismal aging. As telomeres shorten with age, telomere length is considered as a biomarker of aging and a forecaster of longevity [102].

6. Conclusion

The ends of all linear chromosomes face the same difficulties regardless of their structures. Chromosome ends are not replicated completely by the standard replication machinery, resulting in loss of sequence and a 3' overhang on half of the replication products. Early eukaryotes may have solved the end replication problem by co-opting a reverse transcriptase that had arisen in a retrotransposable element [124] and using it to generate arrays of a simple repeat unit. They then solved the end protection problem by engineering long 3' overhangs on all termini, which could then loop around and tuck into the double stranded telomeric region and coating the terminal arrays with proteins that recognize the product of the reverse transcriptase. This combination of telomere maintenance by telomerase and protection by CST/shelterin served eukaryotes well and has been amazingly stable for more than a billion years.

Depending on how strictly shelterin recognizes the telomeric motif, the sequence may be conserved over long expanses of time, as in unikonts. If shelterin is less strict in recognizing the telomerase-generated motif, this sequence may have more latitude to vary, as in chromalveolates. If the protective telomere cap completely loses its ability to recognize the telomeric sequence, telomerase and the canonical telomeric motif may be lost. Many eukaryotes, including yeast and humans use unequal homologous recombination/gene conversion as a backup telomere maintenance system. It appears that in some species of plants and animals telomerase has been lost, and gene conversion has taken over as the primary mechanism to maintain chromosome length, with the eventual loss of telomeric motif. Chromosome length maintenance and end protection must be maintained through all of this. The evolution of new telomere structures, therefore, requires a delicate interplay between these two functions, as well as other telomeric functions that may be less well understood, such as heterochromatin formation and meiotic chromosome pairing.

Acknowledgements

The authors were supported in part by the Intramural Research Program, NIEHS, U. S. National Institutes of Health and by grant no. GAČR P501/10/1215 (Grant Agency of the Czech Republic, Prague).

Author details

Radmila Capkova Frydrychova[1*] and James M. Mason[2*]

*Address all correspondence to: radmila.frydrychova@hotmail.com

1 Institute of Entomology, Czech Republic

2 National Institute of Environmental Health Sciences, USA

References

[1] Greider, C.W., & Blackburn, E. H. (1996). Telomeres, telomerase and cancer. *Sci Am*, 274(2), 92-97.

[2] Mason, J. M., Frydrychova, R. C., & Biessmann, H. (2008). Drosophila telomeres: an exception providing new insights. *Bioessays*, 30(1), 25-37.

[3] Biessmann, H., & Mason, J. M. (2003). Telomerase-independent mechanisms of telomere elongation. *Cell Mol Life Sci*, 60(11), 2325-2333.

[4] Zakian, V. A. (2012). Telomeres: The beginnings and ends of eukaryotic chromosomes. *Exp Cell Res*, 318(12), 1456-1460.

[5] Sohanpal, B. K., Morzaria, S. P., Gobright, E. I., & Bishop, R. P. (1995). Characterisation of the telomeres at opposite ends of a 3 Mb Theileria parva chromosome. *Nucleic Acids Res*, 23(11), 1942-1947.

[6] Liu, C., Schroeder, A. A., Kapur, V., & Abrahamsen, M. S. (1998). Telomeric sequences of Cryptosporidium parvum. *Mol Biochem Parasitol*, 94(2), 291-296.

[7] Klobutcher, L. A., Swanton, M. T., Donini, P., & Prescott, D. M. (1981). All gene-sized DNA molecules in four species of hypotrichs have the same terminal sequence and an unusual 3' terminus. *Proc Natl Acad Sci U S A*, 78(5), 3015-3019.

[8] Alverca, E., Cuadrado, A., Jouve, N., Franca, S., & Moreno Diaz., de la Espina. (2007). Telomeric DNA localization on dinoflagellate chromosomes: structural and evolutionary implications. *Cytogenet Genome Res*, 116(3), 224-231.

[9]   Bowler, C., Allen, A. E., Badger, J. H., Grimwood, J., Jabbari, K., Kuo, A., et al. (2008). The Phaeodactylum genome reveals the evolutionary history of diatom genomes. *Nature*, 456(7219), 239-244.

[10] Koonin, E. V. (2010). Preview .The incredible expanding ancestor of eukaryotes. *Cell*, 140(5), 606-608.

[11] Gilson, P. R., & McFadden, G. I. (2002). Jam packed genomes--a preliminary, comparative analysis of nucleomorphs. *Genetica*, 115(1), 13-28.

[12] Wellinger, R. J., & Sen, D. (1997). The DNA structures at the ends of eukaryotic chromosomes. *Eur J Cancer*, 33(5), 735-749.

[13] Nozaki, H., Takano, H., Misumi, O., Terasawa, K., Matsuzaki, M., & Maruyama, S. (2007). A 100%-complete sequence reveals unusually simple genomic features in the hot-spring red alga Cyanidioschyzon merolae. *BMC Biol*, 5, 28.

[14] Fulneckova, J., Hasikova, T., Fajkus, J., Lukesova, A., Elias, M., & Sykorova, E. (2012). Dynamic evolution of telomeric sequences in the green algal order Chlamydomonadales. *Genome Biol Evol*, 4(3), 248-264.

[15] Pich, U., Fuchs, J., & Schubert, I. (1996). How do Alliaceae stabilize their chromosome ends in the absence of TTTAGGG sequences? *Chromosome Res*, 4(3), 207-213.

[16] Schubert, I., Rieger, R., Fuchs, J., & Pich, U. (1994). Sequence organization and the mechanism of interstitial deletion clustering in a plant genome (Vicia faba). *Mutat Res*, 325(1), 1-5.

[17] Fajkus, J., Kovarik, A., Kralovics, R., & Bezdek, M. (1995). Organization of telomeric and subtelomeric chromatin in the higher plant Nicotiana tabacum. *Mol Gen Genet*, 247(5), 633-638.

[18] Peska, V., Sykorova, E., & Fajkus, J. (2008). Two faces of Solanaceae telomeres: a comparison between Nicotiana and Cestrum telomeres and telomere-binding proteins. *Cytogenet Genome Res*, 122(3-4), 380-387.

[19] Sykorova, E., Lim, K. Y., Fajkus, J., & Leitch, A. R. (2003). The signature of the Cestrum genome suggests an evolutionary response to the loss of (TTTAGGG)n telomeres. *Chromosoma*, 112(4), 164-172.

[20] Watson, J. M., & Riha, K. (2010). Comparative biology of telomeres: where plants stand. *FEBS Lett*, 584(17), 3752-3759.

[21] Vitkova, M., Kral, J., Traut, W., Zrzavy, J., & Marec, F. (2005). The evolutionary origin of insect telomeric repeats, (TTAGG)n. *Chromosome Res*, 13(2), 145-156.

[22] Frydrychova, R., & Marec, F. (2002). Repeated losses of TTAGG telomere repeats in evolution of beetles (Coleoptera). *Genetica*, 115, 179-187.

[23] Frydrychova, R., Grossmann, P., Trubac, P., Vitkova, M., & Marec, F. (2004). Phylogenetic distribution of TTAGG telomeric repeats in insects. *Genome*, 47(1), 163-178.

[24] Mravinac, B., Mestrovic, N., Cavrak, V. V., & Plohl, M. (2011). TCAGG, an alternative telomeric sequence in insects. *Chromosoma*, 120(4), 367-376.

[25] Mason, J. M., & Biessmann, H. (1995). The unusual telomeres of Drosophila. *Trends Genet*, 11(2), 58-62.

[26] Sasaki, T., & Fujiwara, H. (2000). Detection and distribution patterns of telomerase activity in insects. *Eur J Biochem*, 267(10), 3025-3031.

[27] Wiegmann, B. M., Trautwein, M. D., Winkler, I. S., Barr, N. B., Kim, J. W., Lambkin, C., et al. (2011). Episodic radiations in the fly tree of life. *Proc Natl Acad Sci U S A*, 108(14), 5690-5695.

[28] Lundblad, V., & Blackburn, E. H. (1993). An alternative pathway for yeast telomere maintenance rescues est1- senescence. *Cell.*, 73(2), 347-360.

[29] Nielsen, L., & Edstrom, J. E. (1993). Complex telomere-associated repeat units in members of the genus Chironomus evolve from sequences similar to simple telomeric repeats. *Mol Cell Biol*, 13(3), 1583-1589.

[30] Nielsen, L., Schmidt, E. R., & Edstrom, J. E. (1990). Subrepeats result from regional DNA sequence conservation in tandem repeats in Chironomus telomeres. *J Mol Biol*, 216(3), 577-584.

[31] Cohn, M., & Edstrom, J. E. (1992). Chromosome ends in Chironomus pallidivittatus contain different subfamilies of telomere-associated repeats. *Chromosoma*, 101(10), 634-640.

[32] Biessmann, H., Walter, M. F., & Mason, J. M. (1997). Drosophila telomere elongation. *Ciba Found Symp*, 211, 53-67.

[33] Biessmann, H., Kobeski, F., Walter, M. F., Kasravi, A., & Roth, C. W. (1998). DNA organization and length polymorphism at the 2L telomeric region of Anopheles gambiae. *Insect Mol Biol*, 7(1), 83-93.

[34] Roth, C. W., Kobeski, F., Walter, M. F., & Biessmann, H. (1997). Chromosome end elongation by recombination in the mosquito Anopheles gambiae. *Mol Cell Biol*, 17(9), 5176-5183.

[35] Madalena, C. R., Amabis, J. M., & Gorab, E. (2010). Unusually short tandem repeats appear to reach chromosome ends of Rhynchosciara americana (Diptera: Sciaridae). *Chromosoma*, 119(6), 613-623.

[36] Pryde, F. E., Gorham, H. C., & Louis, E. J. (1997). Chromosome ends: all the same under their caps. *Curr Opin Genet Dev*, 7(6), 822-828.

[37] George, J. A., Traverse, K. L., DeBaryshe, P. G., Kelley, K. J., & Pardue, M. L. (2010). Evolution of diverse mechanisms for protecting chromosome ends by Drosophila TART telomere retrotransposons. *Proc Natl Acad Sci U S A*, 107(49), 21052-21057.

[38] Sheen, F. M., & Levis, R. W. (1994). Transposition of the LINE-like retrotransposon TART to Drosophila chromosome termini. *Proc Natl Acad Sci U S A*, 91(26), 12510-12514.

[39] Rashkova, S., Karam, S. E., Kellum, R., & Pardue, M. L. Gag proteins of the two Drosophila telomeric retrotransposons are targeted to chromosome ends. *J Cell Biol*, 159(3), 397-402.

[40] Casacuberta, E., & Pardue, M. L. (2003). HeT-A elements in Drosophila virilis: retrotransposon telomeres are conserved across the Drosophila genus. *Proc Natl Acad Sci U S A*, 100(24), 14091-14096.

[41] Casacuberta, E., & Pardue, M. L. (2003). Transposon telomeres are widely distributed in the Drosophila genus: TART elements in the virilis group. *Proc Natl Acad Sci U S A*, 100(6), 3363-3368.

[42] Greider, C. W. (1999). Telomeres do D-loop-T-loop. *Cell*, 97(4), 419-422.

[43] Tomaska, L., Willcox, S., Slezakova, J., Nosek, J., & Griffith, J. D. (2004). Taz1 binding to a fission yeast model telomere: formation of telomeric loops and higher order structures. *J Biol Chem*, 279(49), 50764-50772.

[44] Muller, H. J., Herskowitz, I. H., Abrahamson, S., & Oster, I. I. (1954). A Nonlinear Relation between X-Ray Dose and Recovered Lethal Mutations in Drosophila. *Genetics*, 39(5), 741-749.

[45] Riha, K., Heacock, M. L., & Shippen, D. E. (2006). The role of the nonhomologous end-joining DNA double-strand break repair pathway in telomere biology. *Annu Rev Genet*, 40, 237-277.

[46] Capper, R, Britt-Compton, B, Tankimanova, M, Rowson, J, Letsolo, B, & Man, S. (2007). The nature of telomere fusion and a definition of the critical telomere length in human cells. *Genes Dev*, 21(19), 2495-2508.

[47] Linger, B. R., & Price, C. M. (2009). Conservation of telomere protein complexes: shuffling through evolution. *Crit Rev Biochem Mol Biol*, 44(6), 434-446.

[48] Palm, W., & de , L. T. (2008). How shelterin protects mammalian telomeres. *Annu Rev Genet*, 42, 301-334.

[49] Chen, L. Y., Liu, D., & Songyang, Z. (2007). Telomere maintenance through spatial control of telomeric proteins. *Mol Cell Biol*, 27(16), 5898-909.

[50] d'Adda di, F. F., Teo, S. H., & Jackson, S. P. (2004). Functional links between telomeres and proteins of the DNA-damage response. *Genes Dev*, 18(15), 1781-1799.

[51] Diotti, R., & Loayza, D. (2011). Shelterin complex and associated factors at human telomeres. *Nucleus*, 2(2), 119-135.

[52] Longhese, M. P. (2008). DNA damage response at functional and dysfunctional telomeres. *Genes Dev*, 22(2), 125-140.

[53] Celli, G. B., Denchi, E. L., & de, L. T. (2006). Ku70 stimulates fusion of dysfunctional telomeres yet protects chromosome ends from homologous recombination. *Nat Cell Biol*, 8(8), 885-890.

[54] Denchi, E. L., & de, L. T. (2007). Protection of telomeres through independent control of ATM and ATR by TRF2 and POT1. *Nature*, 448(7157), 1068-1071.

[55] Lamarche, B. J., Orazio, N. I., & Weitzman, M. D. (2010). The MRN complex in double-strand break repair and telomere maintenance. *FEBS Lett*, 584(17), 3682-3695.

[56] Wu, P., van Rooney, O. M., & de , S. L. T. (2010). Apollo contributes to G overhang maintenance and protects leading-end telomeres. *Mol Cell*, 39(4), 606-617.

[57] Sfeir, A., & de , L. T. (2012). Removal of shelterin reveals the telomere end-protection problem. *Science*, 336(6081), 593-597.

[58] Greider, C. W. (1999). Telomeres do D-loop-T-loop. *Cell*, 97(4), 419-422.

[59] Gu, P, Min, J. N., Wang, Y, Huang, C, Peng, T, & Chai, W. (2012). CTC1 deletion results in defective telomere replication, leading to catastrophic telomere loss and stem cell exhaustion. *EMBO J*, 31(10), 2309-2321.

[60] Raffa, G. D., Raimondo, D., Sorino, C., Cugusi, S., Cenci, G., Cacchione, S., et al. (2010). Verrocchio, a Drosophila OB fold-containing protein, is a component of the terminin telomere-capping complex. *Genes Dev*, 24(15), 1596-1601.

[61] Raffa, G. D., Ciapponi, L., Cenci, G., & Gatti, M. (2011). Terminin: a protein complex that mediates epigenetic maintenance of Drosophila telomeres. *Nucleus*, 2(5), 383-391.

[62] Cenci, G., Siriaco, G., Raffa, G. D., Kellum, R., & Gatti, M. (2003, Jan). The Drosophila HOAP protein is required for telomere capping. *Nat Cell Biol*, 5(1), 82-84.

[63] Raffa, G. D., Siriaco, G, Cugusi, S, Ciapponi, L, Cenci, G, & Wojcik, E. (2009). The Drosophila modigliani (moi) gene encodes a HOAP-interacting protein required for telomere protection. *Proc Natl Acad Sci U S A*, 106(7), 2271-2276.

[64] Komonyi, O., Schauer, T., Papai, G., Deak, P., & Boros, I. M. (2009). A product of the bicistronic Drosophila melanogaster gene CG31241, which also encodes a trimethyl-guanosine synthase, plays a role in telomere protection. *J Cell Sci*, Mar 15, 122(Pt 6), 769-774.

[65] Gao, G., Walser, J. C., Beaucher, M. L., Morciano, P., Wesolowska, N., Chen, J., et al. HipHop interacts with HOAP and HP1 to protect Drosophila telomeres in a sequence-independent manner. *EMBO J*, 29(4), 819-829.

[66] Biessmann, H., Kasravi, B., Bui, T., Fujiwara, G., Champion, L. E., & Mason, J. M. (1994, Apr). Comparison of two active HeT-A retroposons of Drosophila melanogaster. *Chromosoma*, 103(2), 90-98.

[67] Capkova, F. R., Biessmann, H., & Mason, J. M. (2008). Regulation of telomere length in Drosophila. *Cytogenet Genome Res*, 122(3-4), 356-364.

[68] Ciapponi, L., Cenci, G., & Gatti, M. (2006, Jul). The Drosophila Nbs protein functions in multiple pathways for the maintenance of genome stability. *Genetics*, 173(3), 1447-1454.

[69] Ciapponi, L., Cenci, G., Ducau, J., Flores, C., Johnson-Schlitz, D., & Gorski, M. M. (2004). The Drosophila Mre11/Rad50 complex is required to prevent both telomeric fusion and chromosome breakage. Curr Biol Aug 10; , 14(15), 1360-1366.

[70] Oikemus, S. R., McGinnis, N, Queiroz-Machado, J, Tukachinsky, H, Takada, S, Sunkel, C. E., et al. (2004, Aug). Drosophila atm/telomere fusion is required for telomeric localization of HP1 and telomere position effect. *Genes Dev*, 18(15), 1850-1861.

[71] Oikemus, S. R., Queiroz-Machado, J., Lai, K., Mc Ginnis, N., Sunkel, C., & Brodsky, M. H. (2006, May). Epigenetic telomere protection by Drosophila DNA damage response pathways. *PLoS Genet*, 2(5), e71.

[72] Bi, X., Wei, S. C., & Rong, Y. S. Telomere protection without a telomerase; the role of ATM and Mre11 in Drosophila telomere maintenance. *Curr Biol*, 14(15), 1348-1353.

[73] Raffa, G. D., Cenci, G., Siriaco, G., Goldberg, M. L., & Gatti, M. The putative Drosophila transcription factor woc is required to prevent telomeric fusions. *Mol Cell*, 20(6), 821-831.

[74] Cenci, G., Rawson, R. B., Belloni, G., Castrillon, D. H., Tudor, M., Petrucci, R., et al. UbcD1, a Drosophila ubiquitin-conjugating enzyme required for proper telomere behavior. *Genes Dev*, 11(7), 863-875.

[75] Canudas, S, Houghtaling, B. R., Bhanot, M, Sasa, G, Savage, S. A., Bertuch, A. A., et al. A role for heterochromatin protein 1gamma at human telomeres. *Genes Dev*, 25(17), 1807-19.

[76] Iglesias, N., Redon, S., Pfeiffer, V., Dees, M., Lingner, J., & Luke, B. (2011, Jun). Subtelomeric repetitive elements determine TERRA regulation by Rap1/Rif and Rap1/Sir complexes in yeast. *EMBO Rep*, 12(6), 587-593.

[77] Luke, B., & Lingner, J. TERRA: telomeric repeat-containing RNA. *EMBO J*, 28(17), 2503-2510.

[78] Schoeftner, S, & Blasco, M. A. A 'higher order' of telomere regulation: telomere heterochromatin and telomeric RNAs. *EMBO J*, 28(16), 2323-2336.

[79] Biessmann, H., Prasad, S., Semeshin, V. F., Andreyeva, E. N., Nguyen, Q., Walter, M. F., et al. (2005, Dec). Two distinct domains in Drosophila melanogaster telomeres. *Genetics*, 171(4), 1767-1777.

[80] Kurenova, E., Champion, L., Biessmann, H., & Mason, J. M. (1998, Nov). Directional gene silencing induced by a complex subtelomeric satellite from Drosophila. *Chromosoma*, 107(5), 311-320.

[81] Schoeftner, S., & Blasco, M. A. (2010, Apr). Chromatin regulation and non-coding RNAs at mammalian telomeres. *Semin Cell Dev Biol*, 21(2), 186-193.

[82] Su, T. T. Heterochromatin replication: better late than ever. *Curr Biol*, 20(23), R 1018-R1020.

[83] Leach, T. J., Chotkowski, H. L., Wotring, M. G., Dilwith, R. L., & Glaser, R. L. (2000, Sep). Replication of heterochromatin and structure of polytene chromosomes. *Mol Cell Biol*, 20(17), 6308-16.

[84] Gilbert, D. M., Takebayashi, S. I., Ryba, T., Lu, J., Pope, B. D. , Wilson, K. A., et al. (2010). Space and time in the nucleus: developmental control of replication timing and chromosome architecture. *Cold Spring Harb Symp Quant Biol*, 75, 143-153.

[85] Kim, S. M., Dubey, D. D., & Huberman, J. A. Early-replicating heterochromatin. *Genes Dev*, 17(3), 330-335.

[86] Arnoult, N., Schluth-Bolard, C., Letessier, A., Drascovic, I., Bouarich-Bourimi, R., Campisi, J., et al. (2010, Apr). Replication timing of human telomeres is chromosome arm-specific, influenced by subtelomeric structures and connected to nuclear localization. *PLoS Genet*, 6(4), e1000920.

[87] Bianchi, A., & Shore, D. Molecular biology Refined view of the ends. *Science*, 320(5881), 1301-1302.

[88] Bianchi, A., & Shore, D. Early replication of short telomeres in budding yeast. *Cell*, 128(6), 1051-1062.

[89] Stewart, J. A., Chaiken, M. F., Wang, F., & Price, C. M. (2012). Maintaining the end: roles of telomere proteins in end-protection, telomere replication and length regulation. Mutat Res Feb 1 , 730(1-2), 12-19.

[90] Fouche, N., Ozgur, S., Roy, D., & Griffith, J. D. (2006). Replication fork regression in repetitive DNAs. *Nucleic Acids Res*, 34(20), 6044-6050.

[91] Sampathi, S., & Chai, W. (2011, Jan). Telomere replication: poised but puzzling. *J Cell Mol Med*, 15(1), 3-13.

[92] Yang, Q., Zhang, R., Wang, X. W., Spillare, E. A., Linke, S. P., Subramanian, D., et al. (2002). The processing of Holliday junctions by BLM and WRN helicases is regulated by. J Biol Chem Aug 30 , 277(35), 31980-31987.

[93] Sfeir, A., Kosiyatrakul, S. T., Hockemeyer, D., Mac Rae, S. L., Karlseder, J., Schildkraut, C. L., et al. (2009). Mammalian telomeres resemble fragile sites and require TRF1 for efficient replication. CellJul 10 , 138(1), 90-103.

[94] Miller, K. M., Rog, O., & Cooper, J. P. (2006). Semi-conservative DNA replication through telomeres requires Taz1. NatureApr 6 , 440(7085), 824-828.

[95] Gu, P, Min, J. N., Wang, Y, Huang, C, Peng, T, Chai, W, et al. CTC1 deletion results in defective telomere replication, leading to catastrophic telomere loss and stem cell exhaustion. *EMBO J*, 31(10), 2309-2321.

[96] Raghuraman, M. K., Winzeler, E. A., Collingwood, D., Hunt, S., Wodicka, L., Conway, A., et al. (2001). Replication dynamics of the yeast genome. Science Oct 5 , 294(5540), 115-121.

[97] Greider, C. W., & Blackburn, E. H. The telomere terminal transferase of Tetrahymena is a ribonucleoprotein enzyme with two kinds of primer specificity. Cell (1987). Dec 24 , 51(6), 887-98.

[98] Greider, C. W. (2006). Telomerase RNA levels limit the telomere length equilibrium. *Cold Spring Harb Symp Quant Biol*, 71, 225-229.

[99] Greider, C. W., & Blackburn, E. H. (2004). Tracking telomerase. CellJan 23 2 , 116, S 83-S 86.

[100] Greider, C. W. (1994). Mammalian telomere dynamics: healing, fragmentation shortening and stabilization. Curr Opin Genet Dev Apr; , 4(2), 203-211.

[101] Wright, W. E., Piatyszek, M. A., Rainey, W. E., Byrd, W., & Shay, J. W. (1996). Telomerase activity in human germline and embryonic tissues and cells. *Dev Genet*, 18(2), 173-179.

[102] Zhu, H., Belcher, M., & van der Harst, P. (2011, May). Healthy aging and disease: role for telomere biology? *Clin Sci (Lond)*, 120(10), 427-440.

[103] Le , S., Zhu, J. J., Anthony, D. C., Greider, C. W., & Black, P. M. (1998, May). Telomerase activity in human gliomas. *Neurosurgery*, 42(5), 1120-1124.

[104] Cifuentes-Rojas, C., & Shippen, D. E. (2012). Telomerase regulation. Mutat Res Feb 1 , 730(1-2), 20-7.

[105] Nelson, N. D., & Bertuch, A. A. (2012). Dyskeratosis congenita as a disorder of telomere maintenance. Mutat Res Feb 1 , 730(1-2), 43-51.

[106] Armanios, M. Y., Chen, J. J., Cogan, J. D., Alder, J. K., Ingersoll, R. G., Markin, C., et al. (2007). Telomerase mutations in families with idiopathic pulmonary fibrosis. N Engl J Med Mar 29 , 356(13), 1317-1326.

[107] Greenberg, R. A., O'Hagan, R. C., Deng, H., Xiao, Q., Hann, S. R., Adams, R. R., et al. (1999). Telomerase reverse transcriptase gene is a direct target of c-Myc but is not functionally equivalent in cellular transformation. Oncogene Feb 4 , 18(5), 1219-1226.

[108] Kyo, S., Takakura, M., Fujiwara, T., & Inoue, M. (2008, Aug). Understanding and exploiting hTERT promoter regulation for diagnosis and treatment of human cancers. *Cancer Sci*, 99(8), 1528-1538.

[109] Holt, S. E., Wright, W. E., & Shay, J. W. (1997, Apr). Multiple pathways for the regulation of telomerase activity. *Eur J Cancer*, 33(5), 761-766.

[110] Loayza, D., & de, LT. (2003). POT1 as a terminal transducer of TRF1 telomere length control. Nature Jun 26 , 423(6943), 1013-1018.

[111] Wang, F., Podell, E. R., Zaug, A. J., Yang, Y., Baciu, P., Cech, T. R., et al. (2007). The POT1-TPP1 telomere complex is a telomerase processivity factor. Nature Feb 1 , 445(7127), 506-510.

[112] Zaug, A. J., Podell, E. R., Nandakumar, J., & Cech, T. R. (2010). Functional interaction between telomere protein TPP1 and telomerase. Genes Dev Mar 15 , 24(6), 613-22.

[113] Zaug, A. J., Podell, E. R., & Cech, T. R. (2005). Human POT1 disrupts telomeric G-quadruplexes allowing telomerase extension in vitro. Proc Natl Acad Sci U S A Aug 2 , 102(31), 10864-10869.

[114] Savitsky, M., Kravchuk, O., Melnikova, L., & Georgiev, P. (2002, May). Heterochromatin protein 1 is involved in control of telomere elongation in Drosophila melanogaster. *Mol Cell Biol*, 22(9), 3204-3218.

[115] Perrini, B., Piacentini, L., Fanti, L., Altieri, F., Chichiarelli, S., & Berloco, M. (2004). HP1 controls telomere capping, telomere elongation, and telomere silencing by two different mechanisms in Drosophila. Mol Cell Aug 13 , 15(3), 467-476.

[116] Frydrychova, R. C., Biessmann, H., Konev, A. Y., Golubovsky, M. D., Johnson, J., Archer, T. K., et al. (2007, Jul). Transcriptional activity of the telomeric retrotransposon HeT-A in Drosophila melanogaster is stimulated as a consequence of subterminal deficiencies at homologous and nonhomologous telomeres. *Mol Cell Biol*, 27(13), 4991-5001.

[117] Frydrychova, R. C., Mason, J. M., & Archer, T. K. (2008, Sep). HP1 is distributed within distinct chromatin domains at Drosophila telomeres. *Genetics*, 180(1), 121-131.

[118] Torok, T., Benitez, C., Takacs, S., & Biessmann, H. (2007, Apr). The protein encoded by the gene proliferation disrupter (prod) is associated with the telomeric retrotransposon array in Drosophila melanogaster. *Chromosoma*, 116(2), 185-195.

[119] Savitsky, M., Kwon, D., Georgiev, P., Kalmykova, A., & Gvozdev, V. (2006). Telomere elongation is under the control of the RNAi-based mechanism in the Drosophila germline. Genes Dev Feb 1 , 20(3), 345-354.

[120] George, J. A., & Pardue, M. L. (2003, Feb). The promoter of the heterochromatic Drosophila telomeric retrotransposon, HeT-A, is active when moved into euchromatic locations. *Genetics*, 163(2), 625-635.

[121] Walter, M. F., & Biessmann, H. (2004, May). Expression of the telomeric retrotransposon HeT-A in Drosophila melanogaster is correlated with cell proliferation. *Dev Genes Evol*, 214(5), 211-9.

[122] Oikawa, S., Tada-Oikawa, S., & Kawanishi, S. (2001). Site-specific DNA damage at the GGG sequence by UVA involves acceleration of telomere shortening. Biochemistry Apr 17 , 40(15), 4763-8.

[123] Barja, G. (2002). Rate of generation of oxidative stress-related damage and animal longevity. Free Radic Biol Med Nov 1 , 33(9), 1167-1172.

[124] Belfort, M., Curcio, M. J., & Lue, N. F. (2011). Telomerase and retrotransposons: reverse transcriptases that shaped genomes. Proc Natl Acad Sci U S A Dec 20 , 108(51), 20304-20310.

Permissions

The contributors of this book come from diverse backgrounds, making this book a truly international effort. This book will bring forth new frontiers with its revolutionizing research information and detailed analysis of the nascent developments around the world.

We would like to thank David Stuart, for lending his expertise to make the book truly unique. He has played a crucial role in the development of this book. Without his invaluable contribution this book wouldn't have been possible. He has made vital efforts to compile up to date information on the varied aspects of this subject to make this book a valuable addition to the collection of many professionals and students.

This book was conceptualized with the vision of imparting up-to-date information and advanced data in this field. To ensure the same, a matchless editorial board was set up. Every individual on the board went through rigorous rounds of assessment to prove their worth. After which they invested a large part of their time researching and compiling the most relevant data for our readers. Conferences and sessions were held from time to time between the editorial board and the contributing authors to present the data in the most comprehensible form. The editorial team has worked tirelessly to provide valuable and valid information to help people across the globe.

Every chapter published in this book has been scrutinized by our experts. Their significance has been extensively debated. The topics covered herein carry significant findings which will fuel the growth of the discipline. They may even be implemented as practical applications or may be referred to as a beginning point for another development. Chapters in this book were first published by InTech; hereby published with permission under the Creative Commons Attribution License or equivalent.

The editorial board has been involved in producing this book since its inception. They have spent rigorous hours researching and exploring the diverse topics which have resulted in the successful publishing of this book. They have passed on their knowledge of decades through this book. To expedite this challenging task, the publisher supported the team at every step. A small team of assistant editors was also appointed to further simplify the editing procedure and attain best results for the readers.

Our editorial team has been hand-picked from every corner of the world. Their multi-ethnicity adds dynamic inputs to the discussions which result in innovative outcomes. These outcomes are then further discussed with the researchers and contributors who give their valuable feedback and opinion regarding the same. The feedback is then collaborated with the researches and they are edited in a comprehensive manner to aid the understanding of the subject.

Apart from the editorial board, the designing team has also invested a significant amount of their time in understanding the subject and creating the most relevant covers. They scrutinized every image to scout for the most suitable representation of the subject and create an appropriate cover for the book.

The publishing team has been involved in this book since its early stages. They were actively engaged in every process, be it collecting the data, connecting with the contributors or procuring relevant information. The team has been an ardent support to the editorial, designing and production team. Their endless efforts to recruit the best for this project, has resulted in the accomplishment of this book. They are a veteran in the field of academics and their pool of knowledge is as vast as their experience in printing. Their expertise and guidance has proved useful at every step. Their uncompromising quality standards have made this book an exceptional effort. Their encouragement from time to time has been an inspiration for everyone.

The publisher and the editorial board hope that this book will prove to be a valuable piece of knowledge for researchers, students, practitioners and scholars across the globe.

List of Contributors

Takashi Moriyama and Naoki Sato
Department of Life Sciences, Graduate School of Arts and Sciences, The University of Tokyo, Tokyo, Japan
JST, CREST, Gobancho, Chiyoda-ku, Tokyo, Japan

Hervé Seligmann
The National Collections of Natural History at the Hebrew University of Jerusalem, Israel

Takeo Kubota, Kunio Miyake and Takae Hirasawa
Department of Epigenetic Medicine, Faculty of Medicine, University of Yamanashi, Japan

Amine Aloui, Alya El May and Ahmed Landoulsi
Carthage University, Faculty of Sciences of Bizerta Zarzouna, Tunisia

Saloua Kouass Sahbani
Carthage University, Faculty of Sciences of Bizerta Zarzouna, Tunisia
Sherbrooke University, Faculty of Medecine Québec, Canada

María Vittoria Di Tomaso, Pablo Liddle, Laura Lafon-Hughes, Ana Laura Reyes-Ábalos and Gustavo Folle
Department of Genetics, Instituto de Investigaciones Biológicas Clemente Estable, Montevideo, Uruguay

Douglas Maya, Macarena Morillo-Huesca, Lidia Delgado Ramos, Sebastián Chávez and Mari-Cruz Muñoz-Centeno
Department of Genetics, Faculty of Biology, University of Seville, Spain

Angélique Galvani
CNRS, UFIP (FRE 3478), Univ. of Nantes, Epigenetics: Proliferation and Differentiation, F-44322 Nantes, France
CNRS, UMR7216 Epigenetics and Cell Fate, F-75013 Paris, France
Univ Paris Diderot, Sorbonne Paris Cité, F-75013 Paris, France

Christophe Thiriet
CNRS, UFIP (FRE 3478), Univ. of Nantes, Epigenetics: Proliferation and Differentiation, F-44322 Nantes, France

Andrey Grach
Khmelnitsky Regional Hospital, Ukraine

Radmila Capkova Frydrychova
Institute of Entomology, Czech Republic

James M. Mason
National Institute of Environmental Health Sciences, USA